# 国际竹类栽培品种登录报告

# International Cultivar Registration Report for Bamboos

# （2015—2016）

史军义 主编

ICRCB

科学出版社
北京

## 内 容 简 介
**A brief introduction**

本书是国际竹类栽培品种登录中心（2015—2016）的年度工作总结。全书分为两大部分：第一部分是根据国际园艺学会栽培品种登录特别委员会的统一格式和内容要求撰写的国际竹类栽培品种登录权威（ICRA）年度报告中文版；第二部分是同一报告的英文版。每个部分均包含5个方面内容：①新登录的竹类栽培品种；②已发表的竹类栽培品种整理；③与竹栽培品种国际登录有关的纸质出版物；④与竹栽培品种国际登录有关的网站建设；⑤国际竹类栽培品种登录园建设。最后是与前述内容相关的重要参考文献。

This book is the annual work summary (2015-2016) of International Cultivar Registration Center for Bamboos (ICRCB). It covers two parts: ① the Chinese version of the annual report of International Cultivars Registration Authority for Bamboos (ICRA) in the form requested by The Special Commission for Cultivar Registration of International Society for Horticultural Science; ② the English version of the annual report. The both versions include five sections: ① new registered bamboo cultivars; ② revision for the bamboo cultivars published; ③ the paper publications for the bamboo cultivars; ④ websites construction related with international registration for bamboo cultivars; ⑤ international registration gardens for bamboo cultivars. Finally it is listed for the important references relevant to the all contents mentioned above.

---

图书在版编目（CIP）数据

国际竹类栽培品种登录报告：2015—2016 / 史军义主编. —北京：科学出版社，2017
ISBN 978-7-03-055812-1

Ⅰ. ①国… Ⅱ. ①史… Ⅲ. ①竹-品种-世界 Ⅳ. ①S795

中国版本图书馆 CIP 数据核字（2017）第 301014 号

责任编辑：童安齐　王　钰 / 责任校对：刘玉靖
责任印制：吕春珉 / 封面设计：金舵手世纪

**科学出版社** 出版
北京东黄城根北街 16 号
邮政编码：100717
http://www.sciencep.com

**三河市骏杰印刷有限公司** 印刷
科学出版社发行　各地新华书店经销

*

2017年12月第 一 版　　开本：787×1092　1/16
2017年12月第一次印刷　　印张：12 3/4
字数：178 000
定价：98.00 元

（如有印装质量问题，我社负责调换〈骏杰〉）
销售部电话 010-62136230　编辑部电话 010-62137026

**版权所有，侵权必究**
举报电话：010-64030229；010-64034315；13501151303

# 国际竹类栽培品种登录委员会
# International Cultivar Registration Committee for Bamboos

顾　　问：靳晓白
主　　任：史军义
副 主 任：易同培　周德群
执行委员：张玉霄
委　　员：周志华　赵世伟　孙卫邦
　　　　　刘锦超　陈其兵　刘青林
　　　　　马丽莎　孙茂盛　杨　林
　　　　　令狐启霖　李青　王道云
　　　　　刘宇韬　陈双林　吴劲旭
法律顾问：肖登国
秘　　书：史蓉红　姚　俊　李志伟　蒲正宇

**Advisors:** Jin Xiaobai

**Chairman:** Shi Junyi

**Deputy Chairmen:** Yi Tongpei, Zhou Dequn

**Executive Member:** Zhang Yuxiao

**Member:** Zhou Zhihua, Zhao Shiwei, Sun Weibang,
　　　　Liu Jinchao, Chen Qibing, Liu Qinglin,
　　　　Ma Lisha, Sun Maosheng, Yang Lin,
　　　　Linghu Qilin, Li Qing, Wang Daoyun,
　　　　Liu Yutao, Chen Shuanglin, Wu Jinxu

**Legal Consultant:** Xiao Dengguo

**Secretary:** Shi Ronghong, Yao Jun, Li Zhiwei, Pu Zhengyu

# 沉痛悼念易同培教授

国家有突出贡献的优秀专家、著名竹子分类学家、四川农业大学教授易同培先生因病医治无效，于2016年9月27日下午14时20分不幸逝世，享年83岁。

易同培教授1932年12月4日出生，重庆丰都人，汉族，中共党员。1957年7月云南大学林学系森林经营专业毕业，1957年10月—1998年12月在四川省林业学校工作，1998年12月退休。2002年12月—2016年4月受聘兼职中国林业科学研究院西南花卉研究开发中心工作。

易同培教授长期从事竹类教学、科研和社会服务工作；先后发现竹子新属5个、新种250多个；独著、合著《中国植物志》（第九卷第一分册）、《中国竹类图志》《四川植物志》等13部专著，发表学术论文160余篇；曾获省、部级科技进步奖二等奖2项、三等奖5项，获梁希林业科学技术奖1项。

易同培教授深受师生和同事爱戴，将毕身精力献给了祖国的教育和科技事业，他的人品、师德和敬业精神永远是我们学习的榜样。他爱岗敬业、无私奉献的精神，脚踏实地、一丝不苟的作风，严于律己、为人师表的风范，将永远铭记在我们心中。

易同培教授的不幸逝世，使我们失去了一位好同志、好老师。对他的逝世我们表示深切的哀悼！我们要化悲痛为力量，继续易同培同志的未竟事业，为祖国的富强繁荣乃至世界的和平进步，特别是竹类科学研究及其产业的昌盛与发展多做贡献！

易同培教授安息吧！

# Mourning Prof. Tongpei Yi with Deep Grief

Mr. Tongpei Yi, a famous bamboo taxonomist, deputy chairman of International Cultivar Registration Committee for Bamboos, a professor of Sichuan Agricultural University, died at the age of 83, 14:20 of September 27 2016 due to incurable illness.

Prof. Yi was born on December 4, 1932 at Fengdu County, Chongqing City, China. He was a CPC member. In July 1957, Yi graduated from Forest Management Major, Forestry Department of Yunnan University. From October 1957 to December 1998, Yi had worked at Sichuan Forestry School. In December 1998, Yi was retired from Sichuan Forestry School. In December 2002 to April 2016, He was employed as a senior research professor at Flower Research and Development Center of Southwest China, Chinese Academy of Forestry.

Prof. Yi had been worked in bamboo teaching, research and public service for many years. He published five new bamboo genera, more than 250 new species plus 13 books, such as the 1st fascicle, volume nine, *Flora of China*, *Iconographia Bambusoidearum Sinicarum* and more than 160 papers. Prof. Yi was granted with two times of the 2nd and five times of 3rd Scientific and Technological Progress Prize from Sichuan Province and Ministry of Forestry, China respectively. Yi was also endowed with Liang Xi Forestry science and Technology Award.

Prof. Yi was esteemed and respected deeply by his students and colleagues due to his enthusiastic dedication to the country's educational and scientific career with his lifetime energy and vigor. He is our good model forever in the high moral quality, teachers' ethnics and professionalism. We commemorate Prof. Yi forever in our hearts because he devoted wholeheartedly and selflessly to the work with down-to-earth and meticulous style and demeanor of being strict demand on himself as well as a virtue model for others.

The unexpected death of Prof. Yi means we have lost a good colleague and good professor. Herewith we should express our deep condolences for his death. We should turn grief into power and continue his unfinished business for the prosperity of the peace and progress of China and the world, especially the bamboo research with greater contribution to the prosperity and development of bamboo industry! May Prof. Yi rest in peace!

# 前　言

自 2013 年国际园艺学会栽培植物命名特别委员会（ISHS Special Commission for Cultivar Registration）批准成立国际竹类栽培品种登录中心（International Cultivar Registration Center for Bamboos，ICRCB）以来，已经度过了 3 年多时间。

2013—2014 年度，我们完成了 ICRCB 组建与启动，受理并批准了 8 个竹类新品种的国际登录申请，在专业出版物上发表相关学术论文 6 篇，开通了 ICRCB 网站（http://www.bamboo2013.org），建立了中国北京、中国成都和中国都江堰 3 处国际竹类栽培品种登录园。

到 2016 年底，在国际竹类栽培品种登录中心（ICRCB）各位专家和工作人员的共同努力下，先后开展了一系列卓有成效的工作，包括：①组织出版了《国际竹类栽培品种登录报告（2013—2014）》；②受理并批准了 12 个竹类新品种的国际登录申请；③对以往依据《国际植物命名法规》（International Code of Botanical Nomenclature，ICBN）公开发表的箣竹属 *Bambusa* Retz. corr. Schreber、方竹属 *Chimonobambusa* Makino、绿竹属 *Dendrocalamopsis*（Chia & H. L. Fung）Keng f. 等 3 个竹属、按照《国际栽培植物命名法规》（International Code of Nomenclature for Cultivated Plants，ICNCP）的相关规定和要求进行了系统整理，内容涉及 18 种竹类植物的 73 个栽培品种；④在专业出版物上发表相关学术论文 9 篇；⑤按专业水准对国际竹类栽培品种登录中心（ICRCB）网站 http://www.bamboo2013.org 进行了系统维护，并开通了英文网页；⑥在原先已设立 3 处国际竹类栽培品种登录园的基础上，新设立了中国南阳和昆明 2 处国际竹类栽培品种登录园。

此间，国际竹类栽培品种登录中心还派专家，应中国国家林业局宣传中心之邀，于 2015 年 4 月考察了浙江龙岩的竹产业发展状况；应四川农业大学之邀，于2016年5月考察了四川青神的竹产业发展状况；应中国林学会竹子分会之邀，于2016年10月参加了"第十二届中国竹业学术大会"，并做了"国际竹类栽培品种登录的理论与实践"的学术报告。所有这些，都为国际竹类栽培品种登录事业的传播、推动和发展，发挥了积极的建设性作用。

在此，我们怀着无比沉痛的心情，深切悼念著名竹子分类学家易同培教授。他在自己的晚年，以其80多岁高龄，积极投身国际竹类栽培品种登录中心（ICRCB）的申办、成立和建设，并亲自担任ICRCB核心组织者，不遗余力、无私奉献，做出了杰出贡献。2016年4月，当他整理发表了自己一生中最后一个新竹种后，突发重病住进医院；于2016年9月27日下午2时，因病医治无效不幸离开了我们，走完了他与竹相伴的光辉一生。为了缅怀这位令人尊敬的前辈，我们决定将他生前与我们一起编撰的《国际竹类栽培品种登录报告（2015—2016）》书稿，立即加以补充、修改和完善，并付诸出版，以示纪念。

虽然国际竹类栽培品种登录中心（ICRCB）的各项工作仍处于初创阶段，工作条件、工作经费以及专家资源、信息资源等仍很贫乏，但几乎所有参与者均为其崇高价值和重要意义所鼓舞。ICRCB的使命是为世界上所有从事竹业科研、教学、开发、推广、生产、加工、协作、交流的机构、组织、单位、企业或个人提供竹类栽培品种国际登录的指导和帮助，在努力规范世界范围内竹类栽培品种的名称，确保其系统的科学性、表达的准确性和交流的方便性的同时，鼓励各竹子分布国家或地区选育和生产更多、对人类更具价值的竹栽培品种，引导并推动有价值竹资源的科学化、标准化、规模化和国际化发展，从而促进各产竹国竹产区经济和社会的协调发展，让更多人有机会共享竹经济和竹文明的发展成果。

史军义

2016年12月30日

# Preface

It has been more than three years since approval of International Cultivar Registration Center for Bamboos (ICRCB) by International Society for Horticultural Science (ISHS) Special Commission for Cultivar Registration.

Durng 2013-2014, ICRCB had been organized and initiated. We received and approved the eight bamboo cultivars for international registration. Also six papers about bamboo cultivars registration had been published. ICRCB website (http://www.bamboo2013.org) had been opened to the public, and three international registration gardens including the garden 'Beijing, China', 'Chengdu, China' and 'Dujiangyan, China' had been established.

Until the end of 2016, through co-effort by all the experts and workers of ICRCB, more work has been fulfilled as the follows:

1) International Cultivar Registration Report for Bamboos (2013-2014) was published.

2) International registration for 12 new bamboo cultivars were received and approved.

3) Based on International Code of Nomenclature for Cultivated Plants (ICNCP), we revised and reposited the taxa from *Bambusa* Retz. corr. Schreber, *Chimonobambusa* Makino and *Dendrocalamopsis* (Chia & H. L. Fung) Keng f., formerly published based on International Code of Botanical Nomenclature (ICBN), which contain 73 cultivars from 18 bamboo species.

4) Nine papers relevant to the bamboo cultivars registration were published.

5) the systematic maintainence for ICRCB website was implemented, and ICRCB English version was also opened to public.

6) Two new international registration gardens for bamboo cultivars (Nanyang, China and Kunming, China) were built plus to the previous three gardens.

Upon the formal invitation issued by Mass Media Center of the State Forestry Administration, ICRCB sent some experts on a study tour for bamboo industry development at Longyan City, Zhejiang Province. On the invitation of Sichuan Agricultural University, ICRCB experts investigated bamboo industry development at Qingshen County, Sichuan Province. On the invitation by the Bamboo Branch, Bamboo Society of China, ICRCB experts attended '12th Bamboo Conference of China' and presented an oral report on 'Theory and Practice of Bamboo Cultivars Registration'. All these work mentioned above has played a positive and constructive role in the international registration for bamboo cultivars in terms of spread, promotion and development.

Herein with extremely painful feelings, we mourned the late Prof. Tongpei Yi, a famous bamboo taxonomist. At his more than 80 years old age, Prof. Yi had actively involved in application, establishment and construction of ICRCB. As one of the core team members of ICRCB, Yi spared no efforts and selfless dedication, made outstanding contributions to ICRCB. After published the last one more new bamboo species, suddenly he hospitalized due to serious illness. At 14:00pm of September 27, 2016, Prof. Yi passed away from us and his favorite bamboo study which had accompanied with his lifetime. In order to cherish the memory of this respectable predecessor, we decided to timely publish the manuscript of International Cultivar Registration Report for Bamboos (2015-2016), which the late Prof. Yi co-compiled with us, after suppliment, revision and improvement.

Although ICRCB is at the initial stage, a lot of things such as work condition, budget, experts resources and information need to be improved, we all have been encouraged by the lofty value and significance. ICRCB's mission is to provide professional guidance and assistance for the institutions, organizations, enterprises and individuals in bamboo research, education, R&D, production, processing, coordination and communication. With effort to standardize names of the bamboo cultivars around the world and aim at the systematic scientificity, expressional veracity and communicative convenience, we will encourage all countries and regions in the world to breed and produce more new bamboo cultivars to promote the high value bamboo resources in the light of scientificity, large-scale and internationalization in order to upgrade harmonic development in both economy and societies We also hope more and more people living in the planet share the achievements of bamboo economy and civilization.

**Shi Junyi**

Devember 30, 2016

# 目 录

前言
Preface

## 第1部分 竹类 ICRA 报告（2015—2016）（中文）

**国际园艺学会栽培品种登录特别委员会 ICRA 年度报告（2015—2016）（竹类）**⋯⋯⋯ 3
  1 国际登录的新竹品种 ⋯⋯⋯⋯⋯⋯⋯⋯⋯⋯⋯⋯⋯⋯⋯⋯⋯⋯⋯⋯⋯⋯⋯⋯⋯⋯⋯⋯ 5
    1.1 '金殿花竹' ⋯⋯⋯⋯⋯⋯⋯⋯⋯⋯⋯⋯⋯⋯⋯⋯⋯⋯⋯⋯⋯⋯⋯⋯⋯⋯⋯⋯⋯⋯ 5
    1.2 '伴黄1号' ⋯⋯⋯⋯⋯⋯⋯⋯⋯⋯⋯⋯⋯⋯⋯⋯⋯⋯⋯⋯⋯⋯⋯⋯⋯⋯⋯⋯⋯⋯ 8
    1.3 '倬牡1号' ⋯⋯⋯⋯⋯⋯⋯⋯⋯⋯⋯⋯⋯⋯⋯⋯⋯⋯⋯⋯⋯⋯⋯⋯⋯⋯⋯⋯⋯⋯ 10
    1.4 '秋实' ⋯⋯⋯⋯⋯⋯⋯⋯⋯⋯⋯⋯⋯⋯⋯⋯⋯⋯⋯⋯⋯⋯⋯⋯⋯⋯⋯⋯⋯⋯⋯⋯ 13
    1.5 '绮彩' ⋯⋯⋯⋯⋯⋯⋯⋯⋯⋯⋯⋯⋯⋯⋯⋯⋯⋯⋯⋯⋯⋯⋯⋯⋯⋯⋯⋯⋯⋯⋯⋯ 16
    1.6 '竹海硬头黄' ⋯⋯⋯⋯⋯⋯⋯⋯⋯⋯⋯⋯⋯⋯⋯⋯⋯⋯⋯⋯⋯⋯⋯⋯⋯⋯⋯⋯⋯ 18
    1.7 '花叶唐竹' ⋯⋯⋯⋯⋯⋯⋯⋯⋯⋯⋯⋯⋯⋯⋯⋯⋯⋯⋯⋯⋯⋯⋯⋯⋯⋯⋯⋯⋯⋯ 20
    1.8 '美菱' ⋯⋯⋯⋯⋯⋯⋯⋯⋯⋯⋯⋯⋯⋯⋯⋯⋯⋯⋯⋯⋯⋯⋯⋯⋯⋯⋯⋯⋯⋯⋯⋯ 22
    1.9 '矮脚麻' ⋯⋯⋯⋯⋯⋯⋯⋯⋯⋯⋯⋯⋯⋯⋯⋯⋯⋯⋯⋯⋯⋯⋯⋯⋯⋯⋯⋯⋯⋯⋯ 25
    1.10 '绿矮脚' ⋯⋯⋯⋯⋯⋯⋯⋯⋯⋯⋯⋯⋯⋯⋯⋯⋯⋯⋯⋯⋯⋯⋯⋯⋯⋯⋯⋯⋯⋯ 27
    1.11 '条纹刺黑竹' ⋯⋯⋯⋯⋯⋯⋯⋯⋯⋯⋯⋯⋯⋯⋯⋯⋯⋯⋯⋯⋯⋯⋯⋯⋯⋯⋯⋯ 29
    1.12 '川牡竹' ⋯⋯⋯⋯⋯⋯⋯⋯⋯⋯⋯⋯⋯⋯⋯⋯⋯⋯⋯⋯⋯⋯⋯⋯⋯⋯⋯⋯⋯⋯ 32
  2 已发表的竹类栽培品种整理 ⋯⋯⋯⋯⋯⋯⋯⋯⋯⋯⋯⋯⋯⋯⋯⋯⋯⋯⋯⋯⋯⋯⋯⋯⋯ 35
    2.1 箣竹属 *Bambusa* Retz. corr. Schreber ⋯⋯⋯⋯⋯⋯⋯⋯⋯⋯⋯⋯⋯⋯⋯⋯⋯⋯⋯⋯ 35
      （1）箣竹 *Bambusa blumeana* J. A. et J. H. Schult. f. ⋯⋯⋯⋯⋯⋯⋯⋯⋯⋯⋯⋯⋯ 35
      （2）坭箣竹 *Bambusa dissimulator* McClure ⋯⋯⋯⋯⋯⋯⋯⋯⋯⋯⋯⋯⋯⋯⋯⋯⋯ 37
      （3）长枝竹 *Bambusa dolichoclada* Hayata ⋯⋯⋯⋯⋯⋯⋯⋯⋯⋯⋯⋯⋯⋯⋯⋯⋯ 39

（4）大眼竹 *Bambusa eutuldoides* McClure ·················· 41

（5）孝顺竹 *Bambusa multiplex*（Lour.）Raeuschel ex J. A. & J. H. Schult. ·········· 43

（6）撑篙竹 *Bambusa pervariabilis* McClure ·················· 58

（7）硬头黄竹 *Bambusa rigida* Keng et Keng f. ·················· 59

（8）青皮竹 *Bambusa textilis* McClure ·················· 60

（9）马甲竹 *Bambusa tulda* Roxb. ·················· 65

（10）青秆竹 *Bambusa tuldoides* Munro ·················· 66

（11）佛肚竹 *Bambusa ventricosa* McClure ·················· 68

（12）龙头竹 *Bambusa vulgaris* Schrader ex Wendland ·················· 69

2.2 方竹属 *Chimonobambusa* Makino ·················· 72

（1）狭叶方竹 *Chimonobambusa angustifolia* C. D. Chu & C. S. Chao ·········· 72

（2）寒竹 *Chimonobambusa marmorea*（Mitford）Makino ·················· 73

（3）方竹 *Chimonobambusa quadrangularis*（Fenzi）Makino ·················· 75

（4）八月竹 *Chimonobambusa szechuanensis*（Rendle）Keng f. ·················· 80

2.3 绿竹属 *Dendrocalamopsis*（Chia & H. L. Fung）Keng f. ·················· 81

（1）线耳绿竹 *Dendrocalamopsis lineariaurita* Yi et L.Yang ·················· 81

（2）绿竹 *Dendrocalamopsis oldhami*（Munro）Keng f. ·················· 82

3 与竹栽培品种国际登录有关的纸质出版物 ·················· 85

4 与竹栽培品种国际登录有关的网站建设 ·················· 86

    4.1 网站名称与网址 ·················· 86

    4.2 网站语言 ·················· 86

    4.3 网站栏目设置及介绍 ·················· 86

    4.4 网站运行情况 ·················· 87

    4.5 网站服务情况 ·················· 87

5 国际竹类栽培品种登录园建设 ·················· 88

    5.1 国际竹品种（中国·北京）登录园 ·················· 88

    5.2 国际竹品种（中国·成都）登录园 ·················· 88

    5.3 国际竹品种（中国·都江堰）登录园 ·················· 88

    5.4 国际竹品种（中国·昆明）登录园 ·················· 89

    5.5 国际竹品种（中国·南阳）登录园 ·················· 89

# 第 2 部分　竹类 ICRA 报告（2015—2016）（英文）

## INTERNATIONAL SOCIETY FOR HORTICULTURAL SCIENCE ISHS SPECIAL COMMISSION FOR CULTIVAR REGISTRATION ICRA REPORT FOR (2015-2016) ·················· 93

- 1 Newly registered bamboo cultivars ·················· 95
  - 1.1 'Jindian Huazhu' ·················· 95
  - 1.2 'Yanghuang 1' ·················· 98
  - 1.3 'Zhuomu 1' ·················· 101
  - 1.4 'Qiushi' ·················· 104
  - 1.5 'Qicai' ·················· 107
  - 1.6 'Zhuhai Yingtouhuang' ·················· 109
  - 1.7 'Huayetangzhu' ·················· 111
  - 1.8 'Meiling' ·················· 113
  - 1.9 'Aijiaoma' ·················· 116
  - 1.10 'Luaijiao' ·················· 118
  - 1.11 'Lineata' ·················· 120
  - 1.12 'Chuanmuzhu' ·················· 123
- 2 Summarization of the published bamboo cultivars ·················· 126
  - 2.1 *Bambusa* Retz. corr. Schreber ·················· 126
    - (1) *Bambusa blumeana* J. A. et J. H. Schult.F. ·················· 126
    - (2) *Bambusa dissimulator* McClure ·················· 128
    - (3) *Bambusa dolichoclada* Hayata ·················· 130
    - (4) *Bambusa eutuldoides* McClure ·················· 132
    - (5) *Bambusa multiplex* (Lour.) Raeuschel ex J. A. & J. H. Schult. ·················· 134
    - (6) *Bambusa pervariabilis* McClure ·················· 149
    - (7) *Bambusa rigida* Keng et Keng f. ·················· 151
    - (8) *Bambusa textilis* McClure ·················· 152
    - (9) *Bambusa tulda* Roxb. ·················· 156

(10) *Bambusa tuldoides* Munro ·········· 157

(11) *Bambusa ventricosa* McClure ·········· 160

(12) *Bambusa vulgaris* Schrader ex Wendland ·········· 161

2.2 *Chimonobambusa* Makino ·········· 164

(1) *Chimonobambusa angustifolia* C. D. Chu & C. S. Chao ·········· 164

(2) *Chimonobambusa marmorea* (Mitford) Makino ·········· 165

(3) *Chimonobambusa quadrangularis* (Fenzi) Makino ·········· 168

(4) *Chimonobambusa szechuanensis* (Rendle) Keng f. ·········· 173

2.3 *Dendrocalamopsis* (Chia & H. L. Fung) Keng f. ·········· 174

(1) *Dendrocalamopsis lineariaurita* Yi et L.Yang ·········· 174

(2) *Dendrocalamopsis oldhami* (Munro) Keng f. ·········· 175

3 Publications on Bamboo Cultivar Registration ·········· 178

4 Website Construction relevant to international registration for bamboo cultivars ·········· 179

4.1 Name and Website ·········· 179

4.2 Website languages ·········· 179

4.3 Contents of the website ·········· 179

4.4 Website operation ·········· 180

4.5 Website Service ·········· 180

5 Construction of International Cultivar Registration Garden for Bamboos ·········· 182

5.1 International Cultivar Registration Garden for Bamboos (Beijing, China) ·········· 182

5.2 International Cultivar Registration Garden for Bamboos (Chengdu, China) ·········· 182

5.3 International Cultivar Registration Garden for Bamboos (Dujiangyan, China) ·········· 182

5.4 International Cultivar Registration Garden for Bamboos (Kunming, China) ·········· 183

5.5 International Cultivar Registration Garden for Bamboos (Nanyang, China) ·········· 183

主要参考文献（Selected Bibliography）·········· 184

# 第 1 部分

# 竹类 ICRA 报告

（2015—2016）

（中文）

# 国际园艺学会栽培品种登录特别委员会
# ICRA 年度报告（2015—2016）

## （竹类）

史 军 义
国际竹类栽培品种登录中心
2016 年 12 月
E-mail:esjy@163.com
http://www.bamboo2013.org
地址：中国云南省昆明市盘龙区，
　　　中国林业科学研究院资源昆虫研究所
邮编：650216

# 1 国际登录的新竹品种

2015—2016年,国际竹类栽培品种登录权威共接受了12个竹类栽培品种的国际登录。登录竹品种排名以批准时间先后为序。

## 1.1 '金殿花竹'

栽培品种名： *Phyllostachys vivax* 'Jindian Huazhu'
申请人： 孙茂盛、杨志杰、徐红林、易同培、姚俊
申请时间： 2015年01月29日
品种保存地： 中国云南省昆明市金殿风景区
批准时间： 2015年03月25日
国际登录号： WB-001-2015-009

**品种描述：**

散生竹。秆高5~10 m,直径3~8 cm,秆形通直,初为绿色,老秆变为灰绿色,秆从基部到中上部各节间具浅黄色纵条纹,且条纹宽窄不等,数量多,较密集;节间圆筒形,长25~35 cm,秆环隆起,略高于箨环,秆环、箨环下具一圈白粉。地下茎(竹鞭)绿色,具浅黄色条纹。分枝高,1、2主枝,秆下部一分枝,上部两分枝,枝环隆起。箨鞘纸质,早落,背面密被黑褐色斑块和斑点;箨耳及鞘口䍁毛缺失;箨片外翻,带状长披针形,褶曲明显,小枝具叶2、3枚,叶片长10~16 cm,宽1.2~2 cm。笋期4月中下旬(见图1-1)。

'金殿花竹'是由乌哺鸡竹 *Phyllostachys vivax* McClure 另一栽培品种'黄纹竹' *Ph. vivax* 'Huanwenzhu' 人工居群中的变异植株经分离移植、克隆培育而成。适于栽培观赏,笋供食用。

**与近缘分类群的关键区别：**

'金殿花竹'与'黄纹竹'属近缘分类群,二者的关键区别在于：前者秆呈灰绿色,秆节间具浅黄色纵条纹,且与秆的绿色部分对比较弱,斑纹数量多,细密,宽窄不等,全秆分布;后者秆呈亮绿色,秆节间仅在分枝一侧具

栽培居群 Cultivated populations

**图 1-1 '金殿花竹'**

**Fig. 1-1** *Phyllostachys vivax* **'Jindian Huazhu'**

1、2条金黄色纵条纹，斑纹较粗，且与秆的绿色部分对比强烈（见表1-1）。

表1-1 '金殿花竹'与'黄纹竹'的关键特征对比

| 分类群<br>Bamboo taxa | '金殿花竹'<br>'Jindian Huazhu' | '黄纹竹'<br>'Huanwenzhu' |
|---|---|---|
| 秆<br>Culms | 秆呈灰绿色 | 秆呈亮绿色 |
| 条纹颜色<br>Stripe color | 浅黄色，<br>与秆的绿色部分对比较弱 | 金黄色，<br>与秆的绿色部分对比强烈 |
| 条纹数量<br>Number of stripes | 每节多条，细密，<br>全秆分布 | 每节1、2条，较粗，<br>仅在秆分枝一侧 |

## 1.2 '佯黄1号'

**栽培品种名：** *Bambusa changningensis* 'Yanghuang 1'
**申请人：** 易同培、李本祥
**申请时间：** 2015年06月02日
**品种保存地：** 中国四川省宜宾市世纪竹园
**批准时间：** 2015年07月10日
**国际登录号：** WB-001-2015-010
**品种描述：**

丛生竹。秆高15~19.5 m，直径8~10 cm，梢端直立；全秆共41~53节，节间长（15）35~45（60）cm，圆筒形，有时在具芽或分枝处具极短凹槽，深绿色，无毛，平滑，幼时被厚白粉，秆壁厚0.5~1.1（2.5）cm，髓屑状；箨环隆起，初时绿色或紫色，有光泽，无毛；秆环平；节内长1~1.5 cm，无毛，幼时无白粉。秆芽1枚，扁桃形，光亮，上部边缘具缘毛。秆第15~20节（高6~8 m）开始分枝，每节具多数枝条，斜展，长80~160（300）cm，直径（0.2）0.5~1.5 cm。笋墨绿色，秆箨基部常被黄褐色贴生刺毛；箨鞘早落，长度约为节间长度的2/3，革质，先端宽弧形，有时稍不对称，背面基底常具棕色贴生刺毛，内面光亮，边缘上部初时具灰色或棕色纤毛；箨耳不等大，长圆形，紫色，大的一枚约比小者大1倍，长约1.5 cm，宽约0.7 cm，边缘繸毛密生，弯曲，长7~15 mm；箨舌稍弧形，无毛，高4~6 mm，边缘繸毛长2~3 mm；箨片直立，绿色，三角形，基部宽度约为箨鞘先端的3/4，背面无毛，腹面基部具稀疏棕色贴生小刺毛，无缘毛。小枝具叶（4）6~8（9）枚；叶鞘长7~10 cm，绿色，无毛，纵脉纹及上部纵脊明显，无缘毛；叶耳及鞘口繸毛缺失；叶舌截平形，紫色，高约1 mm，边缘有时具缘毛；叶柄长3~5 mm，淡绿色，无毛；叶片线状披针形，纸质，长（16）20~29（32）cm，宽（1.8）2.4~3（3.6）cm，上面绿色，下面淡绿色，先端长渐尖，基部阔楔形，不对称，次脉7~10对，小横脉模糊，边缘具小锯齿。笋期8~9月（见图1-2）。

'佯黄1号'是佯黄竹*Bambusa changningensis* Yi et B. X. Li通过进一步选优、分离、移植、培育而成的竹类新品种，其秆形更高大、粗壮，生物量更丰富，竹材产量更高，单株秆最重可达32.5 kg、枝叶重达12.5 kg，单位面积产量比佯黄竹高出 1/3 以上。该品种可供造纸和编织竹器，也是竹型材、竹材胶合板等的重要原料，还可作观赏竹；竹笋味甜，较细嫩，供鲜食。

竹秆 Bamboo culms

秆箨 Culm-sheath

竹叶 Bamboo leaves

竹笋 Bamboo shoot

图 1-2 '佯黄 1 号'
Fig. 1-2 *Bambusa changningensis* 'Yanghuang 1'

## 1.3 '倬牡 1 号'

栽培品种名：　　*Dendrocalamus mutatus* 'Zhuomu 1'
申请人：　　　　易同培、李本祥
申请时间：　　　2015 年 06 月 12 日
品种保存地：　　中国四川省宜宾市世纪竹园
批准时间：　　　2015 年 07 月 16 日
国际登录号：　　WB-001-2015-011
品种描述：

丛生竹。秆高 18～25 m，直径 14～18 cm，梢部直立或微弯；全秆具 55～65 节，节间长 60～70 cm，基部数节节间长 15～30 cm，圆筒形，平滑，幼时被白粉，无毛，中空，竹壁厚 0.8～2.5 cm，髓呈屑状；箨环初时灰色，以后变为褐色，窄而较薄，无毛；秆环平，基部数节上具气生根；节内高 1.2～1.6 cm，无毛。秆芽压扁状，贴生，边缘具褐色短纤毛。秆的分枝习性较高，即从第 18～25 节、秆高 10～14 m 开始分枝，主枝粗壮，长达 4.5～5 m，侧枝数枚，较短而细。笋淡紫红色，箨片淡黄绿色，外翻；秆箨早落，半椭圆形，厚革质，短于节间，背面上部贴生褐色刺毛，纵脉纹不甚明显，无缘毛；箨耳及鞘口繸毛缺失；箨舌凹弧形，中部稍高起，紫色，高 2～5 mm，边缘具短而扁平繸毛；箨片外翻，三角形，基部背卷，腹面被贴生棕色刺毛，边缘粗糙。小枝着叶 5～7（8）枚；叶鞘无毛，上部纵脊隆起较甚，常带紫色，无缘毛；叶耳及鞘口繸毛缺失；叶舌近截形，紫色，无毛，高约 1.5 mm；叶柄长 8～10 mm，无毛；叶片披针形，绿色，纸质，无毛，长 24～35 cm，宽 4.5～7.5 cm，先端渐尖，基部阔楔形，次脉 10～13 对，小横脉组成长方形，边缘具小锯齿而粗糙。笋期 8 月（见图 1-3）。

'倬牡 1 号'是倬牡竹 *Dendrocalamus mutatus* Yi et B. X. Li 通过进一步选优、分离、移植、培育而成的竹类新品种，其秆形更高大、粗壮，生物量更丰富，竹材产量更高，其秆高达 22 m 以上，直径达 17 cm 以上。该品种属笋材两用竹种，其笋肉细腻、质地脆嫩、微甜、营养丰富，竹材通直，生物量高。

竹丛 Bamboo clump / 竹笋 Bamboo shoot

图 1-3 '倬牡 1 号'

Fig. 1-3 *Dendrocalamus mutatus* 'Zhuomu 1'

竹叶 Bamboo lesves　　气生根 Spine-aerial roots

秆箨 Culm-sheath　　幼秆 Young culms

**图 1-3** （续）
**Fig. 1-3** (continued)

**1.4 '秋实'**

栽培品种名： *Dendrocalamus membranaceus* 'Qiushi'
申请人： 孙茂盛、汤成松、段生彪、姚俊
申请时间： 2015 年 10 月 08 日
品种保存地： 中国云南省昆明市西南林业大学竹园
批准时间： 2015 年 11 月 18 日
国际登录号： WB-001-2015-012

**品种描述：**

丛生竹。秆高 10~18 m，直径 6~10（13）cm，秆形通直；节间圆筒形，节间长 20~30 cm，初被白粉；秆环平，箨环明显隆起；秆从基部第一节开始至 18 节具有数条金黄色宽窄不等的纵条纹；基部一至三节具气根；秆从基部开始分枝；3 主枝。箨鞘革质，早落，靠秆基部的秆箨长于节间，上部的秆箨短于节间，竹笋或箨新鲜时具棕色和绿色纵条纹，秆箨光亮无毛，鞘口截平，箨舌高 5~8 mm，先端呈不规则齿裂，肩部狭窄；箨耳缺或偶尔存在而甚微小，其上具较为密集的棕色绒毛；箨片外翻，窄长形，长 5~33 cm，宽 2.5~4 cm，平直，背面具棕色绒毛，尤其箨片基部。小枝具叶 4~8 枚；叶舌不明显，高约 1 mm；叶耳镰形，具紫色继毛；叶片披针形，长 11.5~22 cm，宽 10~15 mm，基部楔形，两面均具白色绒毛。笋期 6 月下旬至 9 月上旬（见图 1-4）。

'秋实'是由黄竹 *Dendrocalamus membranaceus* Munro 野生居群变异植株经进一步分离移栽、克隆培育而成的竹类新品种。该品种具有较高的观赏价值，适合造园、盆栽、做竹廊、竹篱、竹小品或成片营造竹景观林。

**与近缘分类群的关键区别：**

'秋实'与黄竹另一栽培品种'花秆黄竹' *D. membranaceus* 'Striatus' 属近缘分类群，关键区别在于：前者秆从基部一直到 18 节甚至以上的节间均具多条宽窄不等的金黄色条纹，数量较多，呈明显花秆现象，竹笋或秆箨新鲜时具棕色或绿色条纹，多而明显，秆箨光滑无毛；后者仅秆基部 1、2 节节间具 1、2 条细金黄色条纹，竹笋或秆箨新鲜时棕色和绿色条纹少而不明显，秆明显呈绿色，秆箨被棕色绒毛（见表 1-2）。

竹丛 Bamboo clump　　　　　竹笋 Bamboo shoot

图 1-4 '秋实'

Fig. 1-4 *Dendrocalamus membranaceus* 'Qiushi'

## 表 1-2 '秋实'与'花秆黄竹'的关键特征对比

| 分类群<br>Bamboo taxa | '秋实'<br>'Qiushi' | '花秆黄竹'<br>'Striatus' |
|---|---|---|
| 秆<br>Culm | 具数条宽窄不等的金黄色纵条纹，条纹分布至秆中部以上 | 金黄色纵条纹少，条纹仅见于秆基部 1、2 节 |
| 秆箨<br>Culm-sheaths | 光滑无毛，具明显棕色和绿色条纹 | 被棕色绒毛，无条纹或条纹不明显 |

## 1.5 '绮彩'

**栽培品种名：** *Bambusa lapidea* 'Qicai'
**申请人：** 孙茂盛、姚俊
**申请时间：** 2015 年 11 月 18 日
**品种保存地：** 中国云南省昆明市西南林业大学竹园
**批准时间：** 2015 年 12 月 25 日
**国际登录号：** WB-001-2015-013
**品种描述：**

丛生竹。秆高 8~18 m，直径 6~12 cm，秆形下部通直，上部呈"之"字形；秆壁厚；节间圆筒形，节间长 22~35 cm，初为绿色，后变为黄色；秆从基部开始至中部具有金黄色宽窄不等的纵条纹；基部数节具气根；分枝低，从基部开始分枝；3 主枝。箨鞘厚革质，上部脱落，下部迟落；秆箨短于节间，箨新鲜时具淡黄色纵条纹，尤其竹笋，秆箨光亮无毛；箨片直立，宽卵状三角形，基部圆形收缩，背面无毛，内面贴生向上棕色刺毛；箨舌高 2~5 mm，边缘具毛；箨耳发达，左右不相等卵形具褶皱，其上密被棕色绒毛；箨舌具不规则齿裂；尤其箨片的基部。小枝具叶 4~12 枚，叶片线状披针形，长 11~26 cm，宽 1.5~3 cm，基部圆形，两面均无毛。笋期 6 月下旬至 8 月上旬（见图 1-5）。

栽培居群 Cultivated populations　　竹笋 Bamboo shoot

图 1-5 '绮彩'
Fig. 1-5 *Bambusa lapidea* 'Qicai'

竹秆 Bamboo culms　　　　　　　竹叶 Bamboo leaves

图 1-5 （续）
**Fig. 1-5** (continued)

'绮彩'是由油簕竹 *Bambusa lapidea* McClure 人工居群中的花秆变异植株经进一步分离、移栽、培育而成的竹类新品种，其秆具宽窄不等的黄色纵条纹，秆箨上具明显淡黄色纵条纹。该品种具有较高的观赏价值，适合造园、盆栽、做竹廊、竹小品或成片营造竹景观林。

## 1.6 '竹海硬头黄'

**栽培品种名：** *Bambusa rigida* 'Zhuhai Yingtouhuang'
**申请人：** 陈其兵、江明艳、吕兵洋、姜涛、李念
**申请时间：** 2016 年 03 月 02 日
**品种保存地：** 中国四川省宜宾市长宁县曙光观赏竹园艺场
**批准时间：** 2016 年 04 月 12 日
**国际登录号：** WB-001-2016-014

**品种描述：**

丛生竹。秆高 10～14 m，直径 4～9 cm；节平，新秆被白色蜡粉，无毛，25～35 节，节间长 30～50 cm，秆节最长 60 cm；主枝明显，每小枝具叶 5～12 枚，叶片矩形，长 12～25 cm，宽 2～3 cm，腹面深绿色，无毛，背面粉绿色；其老竹仍有 1～3 个大型休眠芽。笋期 7～9 月（见图 1-6）。

'竹海硬头黄'是由硬头黄竹 *Bambusa rigida* Keng et Keng f. 人工居群中选育出来的优质高产竹子新品种。秆通直，平均胸径 6.5～7.5 cm，最大可以达 9.2 cm，竹材产量高，单株秆重 18 kg，最重可以达 28 kg。单株质量及亩产量比其他品种硬头黄高 30%～50%。可作撑篙、棚架、农具柄等，也是造纸的好材料。竹笋可食用。无性繁殖能力强，竹枝扦插成活率 80% 以上。

该品种不仅可以广泛用于制浆、造纸等，而且适应性强，可以栽植于荒山、"四旁"及庭园，是良好的生态和园林绿化竹种。

**与传统硬头黄竹的关键区别：**

（1）'竹海硬头黄'比传统硬头黄竹适宜种植范围更广，亚热带、南亚热带，海拔 800 m 以下地带均可栽培。

（2）'竹海硬头黄'比传统硬头黄竹秆更高、径更粗、壁更厚，生物量有显著提升，单位面积竹材产量更高。

（3）'竹海硬头黄'比传统硬头黄竹的年出笋率更高、无性繁殖能力更强。

（4）'竹海硬头黄'对土壤要求相对较低，能够耐一定的水湿和低温，抗寒性明显高于硬头黄竹，2008 年冰冻灾害时没有受冻，抗寒性与绵竹 *Lingnania intermedia*（Hsueh et Yi）Yi 相似。

栽培居群 Cultivated populations

秆箨 Culm-sheath　　　　　　　　　　竹叶 Bamboo leaves

图 1-6　'竹海硬头黄'
Fig. 1-6　*Bambusa rigida* 'Zhuhai Yingtouhuang'

## 1.7 '花叶唐竹'

栽培品种名： *Sinobambusa tootsik* 'Huayetangzhu'
申请人： 陈松河、王振忠
申请时间： 2016 年 3 月 10 日
品种保存地： 中国福建省厦门市园林植物园
批准时间： 2016 年 04 月 12 日
国际登录号： WB-001-2016-015
品种描述：

散生竹。秆高 5～12 m，直径 2～6 cm；节间长 30～40（80）cm，初时被白粉，在节下尤密，老秆有纵肋纹，具分枝的一侧扁平并具沟槽；箨环木栓质隆起，开初具紫褐色刚毛；秆环隆起，与箨环同高。秆每节通常分枝 3 枚，有时多达 5～7 枚，枝环很隆起。箨鞘早落，近长方形，先端钝圆，新鲜时绿色，具黄白色纵条纹，箨鞘两边缘的条纹尤其宽大；箨耳卵形至椭圆形，秆先端者常镰形，表面被绒毛或粗糙，继毛波曲，长达 2 cm；箨舌高约 4 mm，拱形，边缘具短纤毛或无毛；箨片绿色，披针形或长披针形，外翻，边缘有稀疏锯齿，边缘略向内收窄后外延。小枝具叶 3～6（9）枚；叶耳不明显，鞘口继毛放射状，长达 15 mm；叶舌高 1～1.5 mm；叶片长 6～22 cm，宽 1～3.5 cm，叶绿色，具有许多宽窄不等的黄白色纵条纹（见图 1-7）。

'花叶唐竹'是由唐竹 *Sinobambusa tootsik*（Sieb.）Makino 人工居群中的花叶变异植株经分离移植、克隆培育出来的优质观赏竹类新品种，其箨鞘新鲜时绿色，具黄白色纵条纹，两边缘的条纹尤其宽大，叶绿色具有许多宽窄不等的黄白色纵条纹。该品种适于在小区、庭院园林中栽培观赏或成片营造竹景观林。

栽培居群 Cultivated populations　　分枝 Branches

竹叶（1）Bamboo leaves（1）　　竹叶（2）Bamboo leaves（2）

竹笋 Bamboo shoots　　秆箨 Culm-sheath　　幼竹 Young bamboos

图 1-7 '花叶唐竹'
Fig. 1-7 *Sinobambusa tootsik* 'Huayetangzhu'

## 1.8 '美菱'

栽培品种名： *Neosinocalamus fangchengensis* 'Meiling'

申请人： 代梅灵、马丽莎、高桂春、姚俊、翟德华

申请时间： 2016 年 06 月 18 日

品种保存地： 国际竹类栽培品种（中国·南阳）登录园

批准时间： 2016 年 08 月 09 日

国际登录号： WB-001-2016-016

品种描述：

丛生竹。秆高 12～14 m，直径 3～4 cm，梢部弯曲常下垂；全秆具 41～43 节，节间长 40～44 cm，基部节间长约 20 cm，圆筒形，初时被白粉，秆中部以下节间具数条宽窄不等的淡黄色纵条纹，秆壁薄，厚 2～4（5）mm；箨环隆起，紫褐色，无毛；秆环平；节内高 2～3 mm，无毛，被白粉。秆芽扁桃形，无毛。秆分枝较高，始于秆中上部，每节具多枚枝条，斜展，主枝长达 1.7 m，直径 4～5 mm，侧枝纤细，直径 1～2 mm。笋浅绿色，被稀疏棕色贴生刺毛，具淡黄色纵条纹。秆箨早落，革质，短于节间，背面被稀疏棕色贴生短刺毛，鲜时亦具淡黄色纵条纹；箨耳缺；箨舌近截平形，初时紫色，高 2～2.5 mm，边缘䍁毛密，扁平，初时紫褐色，长 5～12 mm；箨片线形或线状三角形，长 8～18 cm，外翻。小枝具叶 8～13 枚；叶鞘绿紫色至绿色，无毛；叶耳及鞘口䍁毛缺；叶舌弧形，紫褐色，无毛，高约 1.5 mm；叶柄淡绿色，无毛，长 1.5～2 mm；叶片线状披针形，长 16～25 cm，宽 2.6～4 cm。笋期 7～8 月（见图 1-8）。

'美菱'是由方城慈竹 *Neosinocalamus fangchengensis* Yi et J. Y. Shi 人工栽培居群中的花秆变异植株经进一步分离、移栽、培育而成的竹类新品种，其秆基部节间具数条明显的宽窄不等的淡黄色纵条纹，嫩秆尤明显，竹笋或秆箨新鲜时亦具明显的棕色和绿色纵条纹。该品种具有较高的观赏价值，由于其可适应 -5℃ 的寒冷气候，适合在近似气候条件的北方地区进行造园、盆栽，做竹廊、竹篱、竹小品或成片营造景观竹林，可在中国广大中部地区尤其是汉水流域推广应用，开发潜力巨大。

竹丛 Bamboo clump

图 1-8 '美菱'

Fig. 1-8 *Neosinocalamus fangchengensis* 'Meiling'

竹秆 Bamboo culm

竹叶 Bamboo leaves

分枝 Branches

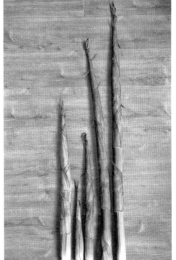
竹笋 Bamboo shoots

图 1-8 （续）
Fig. 1-8 (continued)

## 1.9 '矮脚麻'

栽培品种名： *Dendrocalamus latiflorus* 'Aijiaoma'
申请人： 陈松河、黄克福
申请时间： 2016 年 08 月 01 日
品种保存地： 中国福建省厦门市园林植物园
批准时间： 2016 年 10 月 19 日
国际登录号： WB-001-2016-017

品种描述：

丛生竹。秆高＜12 m，直径＜15 cm，节间长＜30 cm；分枝较低，从第 5 或第 6 节开始分枝，1～3 节具短气生根，枝条排列较整齐；叶片大，纸质，薄而柔软。出笋较早，5 月上旬即开始出笋；笋期较长，每年的 7 月上旬至 8 月上旬为发笋高峰期，11 月上旬停止；单笋重较小，约 1～3 kg，但因笋期长，故单位面积总产量较高（见图 1-9）。

'矮脚麻'是由麻竹 *Dendrocalamus latiflorus* Munro 人工栽培居群中竹笋更适宜食用的植株分离、移栽、培育而成的竹类新品种。该品种竹笋品

竹丛 Bamboo clump

竹秆 Bamboo culm

图 1-9 '矮脚麻'
**Fig. 1-9** *Dendrocalamus latiflorus* 'Aijiaoma'

竹笋 Bamboo shoot　　　　　秆箨 Culm-sheaths

竹叶 Bamboo leaves　　　　　幼竹 Young bamboos

图 1-9 （续）
Fig. 1-9 (continued)

质佳，在闽南地区栽培范围较广，是闽南地区常见的主要优质笋用竹品种之一。因其不甚耐寒，故宜在南亚热带地区推广发展。

## 1.10 '绿矮脚'

栽培品种名： *Dendrocalamopsis oldhami* 'Luaijiao'
申请人： 陈松河、黄克福
申请时间： 2016 年 08 月 01 日
品种保存地： 中国福建省厦门市园林植物园
批准时间： 2016 年 10 月 19 日
国际登录号： WB-001-2016-018
品种描述：

秆丛生。秆高通常 4.0～6.0 m，胸径 4.5～6.0 cm，节间长 30 cm 以下，分枝较低，枝下高 0.8～1.0 m。出笋较早，一般 5 月下旬、个别中旬即开始出笋；笋期较长，每年的 7 月中下旬为发笋高峰期，10 月下旬逐渐停止。笋柄歪斜大，因形似马蹄，俗称"马蹄笋"。其笋体较小，尖削度大，但因笋期长，故单位面积总产量较高（见图 1-10）。

竹丛 Bamboo clump

竹秆 Bamboo culm

图 1-10 '绿矮脚'
Fig. 1-10 *Dendrocalamopsis oldhami* 'Luaijiao'

竹笋 Bamboo shoots

图 1-10 （续）
Fig. 1-10 (continued)

'绿矮脚'是由绿竹 *Dendrocalamopsis oldhami*（Munro）Keng f. 人工栽培居群中竹笋更适宜食用的植株分离、移栽、培育而成的竹类新品种。该品种笋质脆嫩，品质极佳，食用价值高，在闽南地区栽培范围较广，是闽南地区常见的主要优质笋用竹品种之一。竹材篾性较差，但仍可为造纸的原料。因其不甚耐寒，故宜在南亚热带地区推广发展，中亚热带南缘常易受冻，不宜过分北移。

## 1.11 '条纹刺黑竹'

栽培品种名： *Chimonobambusa neopurpurea* 'Lineata'
申请人： 马丽莎、尹显孝、李志伟、刘宇韬、姚俊
申请时间： 2016 年 10 月 06 日
品种保存地： 国际竹类栽培品种（中国·都江堰）登录园
批准时间： 2016 年 10 月 29 日
国际登录号： WB-001-2016-019

品种描述：

秆高 2~5 m，直径 1~3 cm；节间长 10~16（20）cm，绿色，且略呈方形；新秆下部节间为淡紫色，具浅绿色纵条纹；箨环初时密被黄棕色刺毛；秆环稍隆起；节内具发达的气生根刺。秆每节上枝条 3 枚。箨鞘宿存，短于其节间长度；箨耳无，鞘口无继毛，或具少数几条继毛；箨舌圆拱形；箨片直立，锥状，长 1~2 mm。小枝具叶 2~4 枚；叶鞘口无继毛，或有少数几条继毛；叶片线状披针形，长 5~18 cm，宽 0.5~2 cm，下面无毛或有时基部具灰黄色柔毛，次脉 4~6 对，小横脉明显（见图 1-11）。

竹丛 Bamboo Clump

秆箨 Culm-sheaths

图 1-11 '条纹刺黑竹'
Fig. 1-11 *Chimonobambusa neopurpurea* 'Lineata'

竹叶 Bamboo leaves　　　　　　　竹笋 Bamboo shoot

图 1-11 （续）

Fig. 1-11　(continued)

'条纹刺黑竹'是由刺黑竹 *Chimonobambusa neopurpurea* Yi 人工居群中的变异植株经进一步分离移栽、克隆培育而成的竹类新品种。该品种具有较高的观赏价值，适合造园、盆栽或成片营造竹景观林。笋供食用，且品质优良。

**与近缘分类群的关键区别：**

'条纹刺黑竹'与刺黑竹另一栽培品种'都江堰方竹'*Ch. neopurpurea* 'Dujiangyan Fangzhu' 属近缘分类群，二者的关键区别在于：前者其新秆下部节间为淡紫色，具浅绿色纵条纹，秆箨短于其节间长度，笋色相对较深；后者新秆下部节间绿色，无纵条纹，秆箨长于其节间长度，笋色相对较浅（见表 1-3）。

表 1-3 '条纹刺黑竹'与'都江堰方竹'的关键特征对比

| 分类群<br>Bamboo taxa | '条纹刺黑竹'<br>'Lineata' | '都江堰方竹'<br>'Dujiangyan Fangzhu' |
|---|---|---|
| 秆<br>Culms | 新秆下部节间为淡紫色，具浅绿色纵条纹 | 新秆下部节间为绿色，无纵条纹 |
| 秆箨<br>Culm-sheaths | 秆箨短于节间 | 秆箨长于节间 |
| 笋<br>Young shoot | 笋色相对较深 | 笋色相对较浅 |

## 1.12 '川牡竹'

**栽培品种名：** *Dendrocalamus mutatus* 'Chuanmuzhu'
**申请人：** 陈其兵、江明艳、吕兵洋、李念、岑画梦、张成
**申请时间：** 2016 年 11 月 01 日
**品种保存地：** 中国四川省成都市高新区新雅街竹博园
**批准时间：** 2016 年 11 月 15 日
**国际登录号：** WB-001-2016-020

**品种描述：**

'川牡竹'为大型丛生竹，秆高 15~20 m，直径 8~16 cm；秆节处隆起，下部数节有气生根，下部各节均环生短气生根或根点，薄被白粉，幼苗箨环下有一圈较厚的白粉。本种最长节间长度在距地面 5~6 m 处，大多为 45~55 cm，最长者可达 65 cm，向两端逐渐减小；分枝较低，每节多分枝；小枝具叶 6~10 片；叶片呈窄披针形，长 12~30 cm，宽 1.5~3 cm，上表面粗糙，下表面具柔毛，叶缘具小锯齿而粗糙，次脉 3~6 对。箨鞘早落性，厚纸质，先端圆拱形，箨耳微弱。本种属厚壁型竹种，其秆壁最厚的部位是靠近地面的基部，平均壁厚达 4 cm，向上逐渐变薄，平均胸高壁厚为 1.9~2.5 cm，平均秆重 60 kg，最大可以达 86 kg，是竹类植物中秆壁较厚的竹种。笋可食，笋期 7~9 月（见图 1-12）。

'川牡竹'是倬牡竹 *Dendrocalamus mutatus* Yi et B. X. Li 中秆壁较厚的优质植株经分离移植、克隆培育出来的优质材笋两用竹类新品种，同时又可用作大型丛生观赏竹，适于在小区、庭院尤其是风景区园林中栽培观赏或成片营造景观林。

**与近缘分类群的关键区别：**

'川牡竹'与倬牡竹另一栽培品种'倬牡 1 号' *Dendrocalamus mutates* 'Zhuomu 1' 属近缘分类群，二者的关键区别在于：前者秆壁更厚，其平均基部秆壁厚达 4 cm，向上逐渐变薄，平均胸高秆壁厚为 1.9~2.5 cm，平均秆重 60 kg，最大可以达 86 kg，是竹类植物中秆壁较厚的竹种；而后者秆壁相对较薄。

竹丛 Bamboo clump

图 1-12 '川牡竹'

Fig. 1-12 *Dendrocalamus mutatus* 'Chuanmuzhu'

竹秆 Bamboo culms　　　秆箨 Culm-sheaths　　　竹笋 Young shoots

图 1-12　（续）
Fig. 1-12　(continued)

# 2　已发表的竹类栽培品种整理

2015—2016 年，国际竹类栽培品种登录权威对过去已发表的竹亚科簕竹属 *Bambusa* Retz. corr. Schreber、方竹属 *Chimonobambusa* Makino、绿竹属 *Dendrocalamopsis*（Chia & H. L. Fung）Keng f. 进行了比较系统的整理，共涉及 18 种，计 73 品种。其中，簕竹属 12 种、计 55 品种，即簕竹（*Bambusa blumeana*）2 品种、坭簕竹（*Bambusa dissimulator*）3 品种、长枝竹（*Bambusa dolichoclada*）2 品种、大眼竹（*Bambusa eutuldoides*）3 品种、孝顺竹（*Bambusa multiplex*）22 品种、撑篙竹（*Bambusa pervariabilis*）2 品种、硬头黄竹（*Bambusa rigida*）1 品种、青皮竹（*Bambusa textilis*）10 品种、马甲竹（*Bambusa tulda*）1 品种、青秆竹（*Bambusa tuldoides*）3 品种、佛肚竹（*Bambusa ventricosa*）2 品种、龙头竹（*Bambusa vulgaris*）4 品种；方竹属 4 种、14 品种，即狭叶方竹（*Chimonobambusa angustifolia*）1 品种、寒竹（*Chimonobambusa marmorea*）3 品种、方竹（*Chimonobambusa quadrangularis*）8 品种、八月竹（*Chimonobambusa szechuanensis*）2 品种；绿竹属 2 种、4 品种，即线耳绿竹（*Dendrocalamopsis lineariaurita*）1 品种、绿竹（*Dendrocalamopsis oldhami*）3 品种。本次整理的竹栽培品种一律按英文字母排序（原种同名栽培品种优先）。

## 2.1　簕竹属 *Bambusa* Retz. corr. Schreber

### （1）簕竹 *Bambusa blumeana* J. A. et J. H. Schult. f.

秆高 15~24 m，直径 8~15 cm，尾梢下弯，下部略呈"之"字形曲折；节间绿色，长 25~35 cm，幼时于上半部疏被棕色贴生刺毛，老则光滑无毛，秆壁厚 2~3 cm；秆中下部各节均环生短气根或根点，并于箨环之上下方均环生有一圈灰白色或棕色绢毛；分枝常自秆基部第一节开始，下部各节常仅具单枝，且其上的小枝常短缩为弯曲的锐利硬刺，并相互交织而成稠密的刺丛，秆中部和上部各节则为 3 至数枚簇生，主枝显著较粗长。箨鞘迟落，背面密被暗棕色刺毛，干时纵肋隆起，先端作宽拱形或下凹，两侧的顶端各高耸一小尖头；箨耳近相等或稍不相等，线状长圆形，常外翻而呈新月形，边缘密生淡棕色波曲状粗长繸毛；箨舌高 4～5 mm，条裂，边缘被流苏状毛；箨片卵形至狭卵形，常外翻，背面被糙硬毛，腹面密生暗棕色小刺毛，先端渐尖具硬尖头，基部略作圆形收窄后即向两侧平展而成箨耳；箨片基部宽度约为箨鞘先端宽的 2/5，边缘近基部被纤毛。末级小枝具 5～9 叶，叶鞘肋纹隆起，背部的上方被短硬毛，外缘一侧被短纤毛；叶耳微小或缺，鞘口繸毛常不存在或有时仅 2、3 条波曲状短繸毛；叶舌近截形，低矮，边缘微齿裂并被细长纤毛；叶片线状披针形至狭披针形，长 10～20 cm，宽 12～25 mm，两表面均粗糙，近无毛，唯下表面的基部常被稍密的长柔毛，先端渐尖而具粗糙钻状尖头，基部近圆形或近截形。假小穗 2 至数枚簇生于花枝各节；小穗线形，带淡紫色，长 2.5～4 cm，宽 3～4 mm，含小花 4～12 朵，其中 2～5 朵为两性花；颖 2，长约 2 mm，无毛；外稃卵状长圆形，长 6～9 mm，宽 2.5～4 mm，背面无毛，具 9～11 脉，先端尖，边缘无毛；内稃长约 7 mm，宽约 1.8 mm，具 2 脊，脊上密被纤毛，脊间具 3 脉，脊外亦各具 3 脉，先端呈二叉状；花丝分离，长 6～7 mm，花药黄色，宽线形，长 3～4 mm；子房瓶状，长 1.2～2 mm，花柱短，柱头 3 分，羽毛状。笋期 6～9 月，花期春季（但 11 月中旬亦可见开花）。

1) 箣竹 *Bambusa* 'Blumeana'

**又名**：簕竹、郁竹（中国）；Bambu duri（印度尼西亚）；Pring gesing, Shi-chiku（日本）；Buloh duri, Buloh sikai（马来语）；Phai-si-suk, Mai-si-suk（泰国）；Kauayan tinik, Bata-kan, Kawayan-siitan（菲律宾）；Rüssèi roliëk（柬埔寨）；Phaix ba：nz（老挝）；Tre gai（越南）。

**引证**：*Bambusa blumeana* cv. Blumeana, Keng et Wang in Flora Reip. Pop. Sin. 9(1): 53. 1996.

**特征**：与箣竹种特征相似。

**用途**：常栽植为生态防护林；竹秆可作棚架用材。

**分布**：中国（福建，台湾，广西，云南）；印度尼西亚（爪哇岛）和马来西亚东部；菲律宾；泰国；越南；柬埔寨；老挝等。

2) 惠方箣竹 *Bambusa blumeana* 'Wei-fang Lin'

**又名**：惠方簕竹、林氏簕竹（中国）；Rinshi-chiku（日本）

**异名**: *Bambusa blumeana* f. *wei-fang lin*

*Bambusa stenostachya* cv. Wei-fan Lin

**引证**: *Bambusa blumeana* 'Wei-fang Lin', Ohrnb., The Bamb. World. 257. 1999.——*B. blumeana* J. A. et J. H. Schult. cv. Wei-fang Lin (W. C. Lin) Chia et al. in Guihaia 8(2): 123. 1988; Keng et Wang in Flora Reip. Pop. Sin. 9(1): 54. 1996.—— *B. stenostachya* Hack. cv. Wei-fan Lin W. C. Lin in Bull. Taiwan For. Res. Inst. No. 98: 12. f. 7-8, 1964; Fl. Taiwan 5: 761. 1978.——*B. blumeana* J. A. et J. H. Schult. f. *wei-fang lin* (W. C. Lin) Yi in Journ. Sichuan For. Sci. Techn. 28(3): 17. 2007; Yi et al. Icon. Bamb. Sin. 97. 2008, et in Clav. Gen. Spec.Bamb.Sin. 29. 2009; Shi et al. in The Ornamental Bamb. in China. 277. 2012.

**特征**: 与箣竹相似，不同之处在于其秆和分枝的节间均为金黄色，后渐转为橙黄色并具深绿色纵条纹；秆箨淡绿色而具少数乳黄色。常作园林观赏用竹。

**用途**: 作观赏竹，供园林栽培。

**分布**: 中国（台湾北部）。

**（2）坭箣竹 *Bambusa dissimulator* McClure**

秆高 10～18 m，直径 4～7 cm，尾梢近直立或稍下弯，下部略呈"之"字形曲折；节间长 25～35 cm，幼时薄被白蜡粉，通常无毛，有时秆基部数节间可具黄白色纵条纹，秆厚壁；节稍隆起，无毛，偶有基部一、二节生有短气根；分枝常自秆基部第一或第二节就开始分出单枝，上方各节则为 3 至数枝簇生，主枝较粗长，秆下部的分枝上常具硬或软的刺。箨鞘早落，革质，背面近无毛，或被不明显的糙硬毛。干时纵肋隆起，先端呈不对称的拱形；箨耳不相等，常有皱褶，边缘被波曲状继毛，大耳长圆形至倒披针形，宽 4～5 mm，小耳卵形至椭圆形，宽 3～4 mm；箨舌高 5～7 mm，边缘齿裂并条裂，被白色短流苏状毛；箨片直立，卵状三角形至卵状披针形，背面无毛，腹面于脉间被暗棕色小刺毛，基部近圆形或近心形收窄，且其宽度约为箨鞘先端宽的 1/2～3/5，箨片边缘近基部被波曲状继毛。叶鞘通常近无毛，或粗糙或密生短硬毛，但以后变无毛；叶耳不存在或稍微发育，当存在时常为卵形，鞘口继毛少而细弱；叶舌低矮，先端截形，边缘近全缘而无纤毛；叶片线状披针形至披针形，长 7～18 cm，宽 10～18 mm，上表面无毛，下表面疏生短柔毛，尤以沿中脉两侧的毛较密，先端渐尖具钻状尖头，基部近圆形或宽楔形。假小穗单生或以数枚簇生于花枝各节，披针形，形扁，长约 3 cm；先出叶具 2 脊；具芽苞片通常 2 片，卵形，先端钝；小穗轴节间长 2～3 mm，近内稃一侧扁平而无毛，另一侧则粗糙，顶端膨大而其边缘被短纤毛；小穗含两性小花 4 或 5 朵，顶端另有 2 至数朵不孕小花；颖常 1 片或有时不存在，形似外稃而较短；外稃披针形，长达 12 mm，背面无毛，纵脉不明显，先端钝或急尖而具近钻状的短尖头，边缘近顶端有时被短纤毛；内稃具 2 脊，脊于近顶端强折叠，被短纤毛或粗糙，先端常具一小簇毛而近呈画笔状，脊间 5 脉；鳞被 3，近相等，卵形或倒卵形，先端钝或微凹缺，边缘被长纤毛；花丝分离，花药先端钝，微凹；子房倒卵形或卵形，具柄，顶端增厚而被糙硬毛；花柱单一，极短，被毛，柱头 3 分。笋期 7～8 月，花期 3～4 月。

1）坭箣竹 *Bambusa* 'Dissimulator'

**又名**：坭簕竹、簕竹、坭竹、猪姆脯（中国）

**异名**：*Bambusa dissimulator* var. *dissimulator*

**引证**：*Bambusa* 'Dissimulator', Shi et al. in World Bamb. Ratt. 14(6): 27. 2016.——*B. dissimulator* McClure var. *dissimulator*, Keng et Wang in Flora Reip. Pop. Sin. 9(1): 60. 1996.

**特征**：与坭箣竹种特征相似。

**用途**：农村常作绿篱、棚架用材。

**分布**：中国（广东、广西的村庄周围）。

2）白节箣竹 *Bambusa dissimulator* 'Albinodia'

**又名**：白节簕竹（中国）；White Joint Bamboo（英语）

**异名**：*Bambusa dissimulator* var. *albinodia*

**引证**：*Bambusa dissimulator* 'Albinodia', Shi et al. in World Bamb. Ratt. 14(6): 27. 2016.——*B. dissimulator* var. *albinodia* McClure in Lingnan. Sci. Journ. 19(3): 415. 1940; Flora of Guangzhou 774. 1956; Chen, Illustr. manual of Chinese trees and shrubs (supplement). 11. 1957; But et al., Bamboos in Hongkong 31. 1985; Keng et Wang in Flora Reip. Pop. Sin. 9(1): 61. 1996; Ohrnb., The Bamb. World. 259. 1999; Flora of China. 22: 14. 2006; Yi et al. Icon. Bamb. Sin. 99. 2008, et in Clav. Gen. Spec.Bamb.Sin. 31. 2009.

**特征**：与坭簕竹相似，不同之处在于其秆下部各节在箨环上下方均环生一圈灰白色绢毛。

**用途**：农村常作棚架及农作物支柱。

**分布**：中国（广东，香港）。

3）毛簕竹 *Bambusa dissimulator* 'Hispida'

**又名**：毛簕竹（中国）

**异名**：*Bambusa dissimulator* var. *hispida*

**引证**：*Bambusa dissimulator* 'Hispida', Shi et al. in World Bamb. Ratt. 14(6): 27. 2016. ——*B. dissimulator* var. *hispida* McClure in Lingnan Sci. Journ. 19 (3): 415. 1940; Flora of Guangzhou. 774. 1956; Chen, Illustr. manual of Chinese trees and shrubs (supplement) 11. 1957; Keng et Wang in Flora Reip. Pop. Sin. 9(1): 61. 1996; Ohrnb., The Bamb. World. 259. 1999; Flora of China. 22: 14. 2006; Yi et al. in Icon. Bamb. Sin. 99. 2008, et in Clav. Gen. Spec.Bamb. Sin. 31. 2009.

**特征**：与坭簕竹相似，不同之处在于其秆的节和节间以及秆箨的箨鞘背面均明显被糙硬毛。

**用途**：农村常作棚架及农作物支柱。

**分布**：中国（广东）。

**（3）长枝竹 *Bambusa dolichoclada* Hayata**

秆高 10～15 m，直径 4.5～8 cm，尾梢略弯，下部近挺直；节间长 30～45 cm，幼时厚被白蜡粉，无毛，秆壁稍厚；节处不隆起，秆下部数节于箨环之上方环生一圈灰白色绢毛；分枝习性低，常自秆基部第一节开始，以3枝乃至多枝簇生，主枝较粗长。箨鞘早落，革质，背面薄被白蜡粉，并沿顶端及两侧的上半部密生棕色短硬毛，先端稍向外侧的一边倾斜而呈稍不对称的宽拱形，或有时近于斜截形，内侧的边缘上部则被短纤毛；箨耳稍皱缩，其末端近钝圆，边缘及内面均被密生的波曲状继毛，两耳显著不相等，大耳卵状长圆形或狭卵形，横长 2～2.5 cm，高 0.8～1 cm，小耳卵形或椭圆形，其大小约为大耳的 1/3；箨舌高 3～4 mm，边缘微齿裂，并被长达 5 mm 的流苏状毛；箨片直立，易脱落，呈不对称的卵状三角形，背面疏生暗棕色小刺毛，腹面在脉间密生淡棕色小刺毛，先端渐尖具硬尖头，基部略作圆形，收窄后即与两侧箨耳相连，此相连部分为 3~5 mm，箨片基部宽度约为箨鞘先端宽的 2/3，仅在边缘下部被纤毛。叶鞘纵肋明显，被粗硬毛，背部具脊；叶耳小，斜卵形，边缘具长达 8 mm 的继毛；叶舌近截形，极低矮，背面被微糙毛；叶片线形至线状披针形，长 10~26 cm，宽 1~2.3 cm，上表面无毛而显光泽，下表面被短柔毛，先端渐尖具粗糙钻状尖头，基部近钝圆或楔形。假小穗以 3~9 枚簇生于花枝各节；小穗线形，长 3~4 cm，宽 6~8 mm，含小花 4~12 朵，基部托以数片具芽苞片；颖 2 片，卵形或宽卵形，长 2~4.5 mm，具 14 脉，先端急尖；外稃卵形，长约 9 mm，具 18~20 脉，脉间还有小横脉，先端急尖；内稃长约 8.5 mm，具 2 脊，脊上密生短纤毛；花药黄色，长 4.5 mm，顶端微凹缺；子房倒卵形，长 2 mm，顶部疏生短硬毛，花柱极短，柱头 3，羽毛状。

1）长枝竹 *Bambusa* 'Dolichoclada'

**又名**：桶仔竹（中国）；Choshi-chiku（日本）；Long-Branch Bamboo, Long-Shoot Bamboo（英语）

**引证**：*Bambusa* 'Dolichoclada', Shi et al. in World Bamb. Ratt. 14(6): 27. 2016.——*B. dolichoclada* cv. Dolichoclada Hayata, Keng et Wang in Flora Reip. Pop. Sin. 9(1): 94. 1996.

**特征**：与长枝竹种特征相似。

**用途**：生态防护；秆供建筑和制造家具之用；劈篾用以编制农用竹器，如竹篓、米筛、粪箕、斗笠以及捕鱼笼等。

**分布**：中国（福建，台湾）；日本（冲绳、九州）。

2）条纹长枝竹 *Bambusa dolichoclada* 'Stripe'

**异名**：*Bambusa dolichoclada* f. *stripe*

*Bambusa ventricosa* f. *stripe*

**引证**：*Bambusa dolichoclada* 'Stripe', Ohrnb., The Bamb. World. 260. 1999; American Bamboo Society. *Bamboo Species Source List* No. 33: 7. Spring 2013.——*B. dolichoclada* Hayata cv. Stripe W. C. Lin in Bull. Taiwan For. Res. Inst. No.98: 15. f. 9, 10. 1964; Fl. Taiwan 5: 751. 1978; Keng et Wang in Flora Reip. Pop. Sin. 9(1): 96. 1996.——*B. ventricosa* Hayata f. *stripe* (W. C. Lin) Yi; Yi et al. Icon. Bamb. Sin. 118. 2008, et in Clav. Gen.Spec. Bamb.Sin. 37. 2009;

Shi et al. in The Ornamental Bamb. in China. 288. 2012.

**特征**：与长枝竹相似，不同之处在于其秆金黄色，节间在分枝一侧及周围具少数绿色纵条纹；叶主要为绿色，但部分叶片具淡黄白色纵条纹。

**用途**：常用作盆栽、盆景、花坛、小区丛植，以供观赏。

**分布**：中国南方及台湾；日本（九州）。

### （4）大眼竹 *Bambusa eutuldoides* McClure

秆高 6~12 m，直径 4~6 cm，尾梢略弯，下部挺直；节间长 30~40 cm，幼时薄被白蜡粉或近于无粉，无毛或有时仅于上半部疏生脱落性小刺毛；秆壁厚 5 mm，节处稍有隆起，秆基部数节于箨环之上下方各环生一圈灰白色绢毛；分枝常自秆基部第二或第三节开始，以数枝乃至多枝簇生，其中 3 枝较为粗长。箨鞘早落，革质，背面通常无毛，或有时被极稀疏的脱落性贴生小刺毛，干时纵肋稍有隆起，近外侧边缘一边有时具数条纵向黄白色细条纹，先端向外侧一边长下斜，呈极不对称的拱形；箨耳极不相等，形状各异，质极脆，略皱，边缘被波曲状细刚毛，大耳极下延，其下延程度可达箨鞘全长的 2/5~1/2，倒披针形至狭长圆形，长 5~6.5 cm，宽约 1.5 cm，小耳近圆形或长圆形，直径约 1 cm，或有时完全与箨片基部相连，故难以区分；箨舌高 3~5 mm，边缘呈不规则齿裂或条裂，被短流苏状毛；箨片直立，易脱落，呈不对称的三角形至狭三角形，背面疏生脱落性小刺毛，腹面近基部脉间被棕色小刺毛而上部粗糙，先端渐尖具锐利硬尖头，基部略微收窄后即向两侧外延与箨耳相连，此相连部分约为 1 cm，箨片基部宽度约为箨鞘先端宽的 3/5。叶鞘无毛，背部具脊，纵肋隆起；叶耳有时不存在，存在时则呈卵形或狭倒卵形乃至倒卵状长圆形，边缘具数条直或曲的细长繸毛；叶舌高约 0.5 mm，截形，边缘具微齿；叶片披针形至宽披针形，一般长 12~25 cm，宽 14~25 mm，上表面无毛，下表面密生短柔毛，先端骤渐尖具粗糙钻状尖头，基部近圆形或楔形。假小穗以数至多枚簇生花枝各节，线形，长 2.5~5.5 cm；小穗含小花 5、6 朵，基部承托以数片具芽苞片；小穗轴节间形扁，长 3~4 mm，顶端膨大且在其边缘被短纤毛；颖仅 1 片，长圆形，长 9~10 mm，背面无毛，但有许多极小的紫色斑点，具 11 脉，先端急尖具短尖头；外稃与颖相似，但长 12~13 mm，具 13~15 脉；内稃披针形，长约 11 mm，具 2 脊，脊近顶端被短纤毛，脊间 4 脉，脊外每边各具 2

脉；鳞被 3，不相等，边缘被长纤毛，前方 2 片较窄，长约 2 mm，后方 1 片较大，宽倒卵形或近圆形，长约 2 mm；花药长 5 mm，顶端又开；子房近球状，直径约 1 mm，顶端被短硬毛，花柱极短，被短硬毛，柱头 3，羽毛状。颖果幼时近倒卵状，长约 5 mm，顶部被短硬毛，并留有柱头残余。

1）大眼竹 *Bambusa* 'Eutuldoides'

又名：Dai Ngan Bamboo（英语）

异名：*Bambusa eutuldoides* var. *eutuldoides*

引证：*Bambusa* 'Eutuldoides', Shi et al. in World Bamb. Ratt. 14(6): 27. 2016.——*B. eutuldoides* var. *eutuldoides*, Keng et Wang in Flora Reip. Pop. Sin. 9(1): 84. 1996.

特征：与大眼竹种特征相似。

用途：秆可作农村建筑辅助用材，制农具，劈篾用以编结粗竹器等。

分布：中国（广东，广西，香港）。

2）银丝大眼竹 *Bambusa eutuldoides* 'Basistriata'

又名：斑坭竹（中国）

异名：*Bambusa eutuldoides* var. *basistriata*

引证：*Bambusa eutuldoides* 'Basistriata', Shi et al. in World Bamb. Ratt. 14(6): 27. 2016.——*B. eutuldoides* McClure var. *basistriata* McClure in Lingnan Univ. Sci. Bull. No. 9: 9. 1940; Bamboos in Guangxi and cultivation, 36, 1987; Keng et Wang in Flora Reip. Pop. Sin. 9(1): 85. 1996; Ohrnb., The Bamb. World. 261. 1999; Flora of China 22: 22. 2006; Yi et al. Icon. Bamb. Sin. 121. 2008, et in Clav. Gen.Spec. Bamb.Sin. 36. 2009.

特征：与大眼竹特征相似，主要区分在于其秆下部各节间和新鲜箨鞘的背面均为绿色而具黄白色纵条纹，箨耳大的那一枚强波状皱褶。

用途：作观赏竹，供园林栽培。

分布：中国（广西，广东）。

3）青丝黄竹 *Bambusa eutuldoides* 'Viridi-vittata'

异名：*Bambusa eutuldoides* var. *viridi-vittata*

*Bambusa eutuldoides* var. *viridivittata*

引证：*Bambusa eutuldoides* 'Viridi-vittata', American Bamboo Society.

*Bamboo Species Source List* No. 33: 7. Spring 2013.——*B. eutuldoides* McClure var. *viridi-vittata* McClure in Lingnan Univ. Sci. Bull. No. 9: 9. 1940; Bamboos in Guangxi and cultivation, 36, 1987; Keng et Wang in Flora Reip. Pop. Sin. 9(1): 85. 1996. ——*B. eutuldoides* var. *viridivittata*, Ohrnb., The Bamb. World. 261. 1999; Flora of China 22: 22. 2006; Yi et al. Icon. Bamb. Sin. 121. 2008, et in Clav. Gen.Spec. Bamb.Sin. 37. 2009; Shi et al. in The Ornamental Bamb. in China. 285. 2012.

**特征**：与大眼竹特征相似，主要区分在于其秆节间柠檬黄色具绿色纵条纹，箨鞘新鲜时为绿色具柠檬黄色纵条纹，箨耳大的那一枚较大眼竹的为短，强波状皱褶。

**用途**：作观赏竹，供园林栽培。

**分布**：中国（广东，江西，四川）。

**（5）孝顺竹 *Bambusa multiplex*（Lour.）Raeuschel ex J. A. & J. H. Schult.**

秆高 4~6（7）m，直径 2~4 cm；节间绿色，长 30~50 cm，幼时薄被白粉，上半部被棕色小刺毛，秆壁较薄；节稍隆起，无毛。秆分枝始于第二至第三节，各节多枝簇生，主枝稍粗长。箨鞘迟落，背面初时薄被白粉，无毛，先端不对称的拱形；箨耳很小或不明显，边缘具少量繸毛；箨舌高 1~1.5 mm，边缘不规则短齿裂；箨片直立，易脱落，长三角形，背面被暗棕色小刺毛，腹面粗糙，基部宽度与箨鞘顶端近相等。小枝具叶 5~12；叶耳肾形，边缘具波曲细长繸毛；叶舌高约 0.5 mm，边缘微齿裂；叶片长 5~16 cm，宽 0.7~1.6 cm，下面粉绿色，密被灰白色短柔毛。假小穗单生或以数枝簇生于花枝各节，并在基部托有鞘状苞片，线形至线状披针形，长 3~6 cm；先出叶长 3.5 mm，具 2 脊，脊上被短纤毛；具芽苞片通常 1 或 2 片，卵形至狭卵形，长 4~7.5 mm，无毛，具 9~13 脉，先端钝或急尖；小穗含小花（3）5~13，中间小花为两性；小穗轴节间形扁，长 4~4.5 mm，无毛；颖不存在；外稃两侧稍不对称，长圆状披针形，长 18 mm，无毛，具 19~21 脉，先端急尖；内稃线形，长 14~16 mm，具 2 脊，脊上被短纤毛，脊间 6 脉，脊外有一边具 4 脉，另一边具 3 脉，先端两侧各伸出 1 被毛的细长尖头，顶端近截平而边缘被短纤毛；鳞被中两侧的 2 片呈半卵形，长 2.5~3 mm，后方的 1 片细长披针形，长 3~5 mm，边缘无毛；花丝长 8~

10 mm，花药紫色，长 6 mm，先端具一簇白色画笔状毛；子房卵球形，长约 1 mm，顶端增粗而被短硬毛，基部具一长约 1 mm 的子房柄，柱头 3 或其数目有变化，直接从子房顶端伸出，长 5 mm，羽毛状。

1）孝顺竹 *Bambusa* 'Multiplex'

**又名**：火吹竹，火管竹，火广竹，火开竹（中国）；Hourai-chiku, Houou-chiku（日本）；Buloh Cina, Buloh pagar（马来语）；Mai-liang, Mai-phai-lieng（泰国）；Bambu cina（印度尼西亚）；Kawayan tsina, Kawayan sa sonsong（菲律宾）；Cay hop（越南）；Hedge Bamboo（英语）

**异名**：*Bambusa multiplex* var. *multiplex*

**引证**：*Bambusa* 'Multiplex', Shi et al. in World Bamb. Ratt. 14(6): 27. 2016. ——*B. multiplex* var. *multiplex*, Keng et Wang in Flora Reip. Pop. Sin. 9(1): 109. 1996.

**特征**：与孝顺竹种特征相似。

**用途**：分布最广的丛生观赏竹，供园林栽培或编制绿篱。

**分布**：中国（华南至西南部，台湾）；越南；泰国；印度尼西亚；菲律宾。

2）阿博黄纹竹 *Bambusa multiplex* 'Albostriata'

**又名**：Fuiri-houou（日本）；Silverstripe Fernleaf Hedge Bamboo（英语）

**异名**：*Bambusa glaucescens* f. *albosttiata*

*Barnbusa multiplex* f. *albostriata*

*Bambusa multiplex* var. *multiplex*

**引证**：*Bambusa multiplex* 'Albostriata', Ohrnb., The Bamb. World. 268. 1999. ——*B. multiplex* f. *albostriata* Muroi & Sugimoto ex Muroi in J. Himeji Gakuin Wom. Coll. No. 1, 1974:1.——*B. glaucescens* f. *albosttiata* Muroi & Sugimoto. 9,1971——*B. multiplex* var. *multiplex*, Keng et Wang in Flora Reip. Pop. Sin. 9(1): 109. 1996.

**特征**：与孝顺竹特征相似，主要区分在于其秆基部绿色具少数黄白色纵条纹。

**用途**：园艺栽培供观赏。

**分布**：日本（中部）。

3）小琴丝竹 *Bambusa multiplex* 'Alphonse-Karr'

**又名**：Suhou-chiku（日语）；Alphonse Karr Hedge Bamboo（英语）

**异名**：*Bambusa alphonso-Karri*

*Barnbusa alphonso-karrii*

*Bambusa glaucescens* 'Alphonse Karr'

*Barnbusa glaucescens* 'Alphonso-Karrir'

*Barnbusa glaucescens* f. *alphonso-karrii*

*Bambusa glaucescens* f. *alphonso-karri*

*Bambusa multiplex* 'Alphonso-Karr'

*Bambusa multiplex* 'Alphonso-Karri'

*Bambusa multiplex* 'Alphonso-Karrii'

*Bambusa multiplex* f. *alphonso-Karri*

*Barnbusa multiplex* f. *alphonso-karrii*

*Bambusa multiplex* var. *normalis*

*Barnbusa multiplex* var. *normalis* f. *alphonso-karrii*

*Bambusa nana* var. *normalis* f. *alphonso-karrii*

*Bambusa nana* f. *alphonso-karri*

*Bambusa nana* var. *alphonso karri*

*Barnbusa nana* var. *alphonso-karrii*

*Bambusa nana* var. *normalis*

*Leleba multiplex* f. *alphonso-karri*

*Leleba multiplex* f. *alphonso-karrii*

**引证**：*Bambusa multiplex* 'Alphonse-Karr', American Bamboo Society. *Bamboo Species Source List* No. 33: 8. Spring 2013.——*B. multiplex* (Lour.) Raeusch. cv. Alphonse-Karr R. A. Young in USDA Agr. Handb. No.193: 40. 1961; Keng et Wang in Flora Reip. Pop. Sin. 9(1): 112. 1996. ——*B. multiplex* 'Alphonso-Karrii', Ohrnb., The Bamb. World. 267. 1999. —— *B. multiplex* f. *alphonso-karri* (Satow) Nakai in Rika kyoiku 15: 67. 1932; Yi et al. in Icon. Bamb. Sin. 128. 2008, et in Clav. Gen.Spec. Bamb. Sin. 39. 2009; Shi et al. in The Ornamental Bamb. in China. 290. 2012. ——*B. alphonso-Karri* Mitf. ex Satow in Trans Asiat.

Soc. Jap. 27: 91. pl. 3. 1899.; 竹内叔雄，竹的研究（中译本）. 110. 1957. ——*B. alphonso-karrii* Mitford, Bamb. Gard.,: 55, 216. 1896. ——*B. nana* var. *alphonsokarri* (Satow) Marliac ex E. G. Camus, Bambus. 121. 1913. ——*B. nana* var. *norrnalis* f. *alphonso-karrii* Makino in S. Honda, Descr. Prod. For. Jap., 1900: 37.——*B. nana* f. *alphonso-karrii* (Mitford ex Satow) Makino ex Kawamura, 1907: 2. ——*B. nana* Roxb. var. *normalis* Makino ex Shirosawa f. *alphonso-karri* (Mitf. ex Satow) Makino ex Shirosawa, Icon. Bamb. Jap. 56. pl. 9. 1912. ——*B. multiplex* var. *normalis* Sasaki f. *alphonso-karri* Sasaki, Cat. Gov. Herb. (Form.) 68. 1930.——*Leleba multiplex* (Lour.) Nakai f. *alphonso-karri* (Satow) Nakai in Journ. Jap. Bot. 9: 14. 1933. ——*Bambusa multiplex* f. *alphonso-Karri* (Mitf.) Sasaki ex Keng f. in Techn. Bull. Nat'l. For. Res. Bur. China No. 8: 17. 1948; Flora Illustr. Plant. Prima. Sinica. Gramineae. 57, pl. 39. 1959; Bamboos in Guangxi and cultivation, 40, pl. 22. 1987. ——*B. multiplex* 'Alphonse Karr'; R. A. Young in Nation. Hort. Mag. 25, 1946: 260, 264. ——*B. glaucescens* (Lam.) Munro ex Merr. f. *alphonso-karri* (Satow) Hatusima, Fl. Ryukyus .854. 1971. ——*B. glaucescens* 'Alphonso-Karrir'; Hatusima, Woody Pl. Jap., 1976: 316. ——*B. glaucescens* (Willd.) Sieb. ex Munro cv. Alphonse Karr (Young) Chia et But in Photologia 52(4): 258. 1982. —— *B. glaucescens* 'Alphonse Karr'; Crouzet, 1981: 51. ——*B. glaucescens* f. *alphonso-karri* (Mitf.) Wen in Journ. Bamb. Res. 4(2): 16. 1985.

**特征**：与孝顺竹特征相似，主要区分在于其秆和分枝的节间黄色，色泽鲜明，具不同宽度的绿色纵条纹，秆箨新鲜时绿色，具黄白色纵条纹。叶偶尔有几条黄白色条纹。抗冻，可耐−10℃。

**用途**：同孝顺竹，但更具观赏价值，适于公园、小区、庭院栽培观赏。

**分布**：中国（四川，广东，台湾）；日本；欧洲；美国（佛罗里达州）；几乎所有热带国家（南亚、东南亚、东亚）均有栽培。

4）凤尾竹 *Bambusa multiplex* 'Fernleaf'

**异名**：*Bambusa floribunda*

*Bambusa glaucescens* cv. Fernleaf

*Bambusa multiplex* f. *fernleaf*

*Bambusa multiplex* var. *elegans*

*Bambusa multipex* var. *fernleaf*

*Bambusa multiplex* var. *nana*

*Bambusa nana* var. *gracillima*

*Ischurochloa floribunda*

*Leleba elegans*

*Leleba floribunda*

**引证**：*Bambusa multiplex* 'Fernleaf', American Bamboo Society. *Bamboo Species Source List* No. 33: 8. Spring 2013.——B. *multiplex* (Lour.) Raeusch. cv. Fernleaf R. A. Young in USDA Agr. Handb. No. 193: 40. 1961; Fl. Taiwan. 5: 755. 1978; Keng et Wang in Flora Reip. Pop. Sin. 9(1): 113. 1996.——*Ischurochloa floribunda* Buse ex Miq., Fl. Jungh. 390. 1851. —— *Bambusa floribunda* (Buse) Zoll. et Maur. ex Steud., Syn. Pl. Glum. 1: 330. 1854. ——*B. nana* Roxb. var. *gracillima* Makino ex E. G. Camus, Bambus. 121. 1913, non Kurz 1866. ——*Leleba floribunda* (Buse) Nakai in Journ. Jap. Bot. 9: 10. pl. 1. 1933. ——*L. elegans* Koidz. in Act. Phytotax. Geobot. 3: 27. 1934.——*Bambusa multipex* var. *fernleaf* R. A. Yung in Nat'l. Hort. Mag. 25: 261. 1946. ——*B. multiplex* var. *nana* (Roxb.) Keng f. in Techn. Bull. Nat'l. For. Res. Bur. China No.8: 17. 1948, non *B. nana* Roxb. 1832; Flora Illustr. Plant. Prima. Sinica. Gramineae. 57, pl. 38. 1959; Bamboos in Guangxi and cultivation, 41, pl. 23, 1987. —— *B. multiplex* var. *elegans* (Koidz.) Muroi ex Sugimoto, New Keys Jap. Trees. 457. 1961; S. Suzuki. Ind. Jap. Bambusas. 104, 105 (pl.18), 340. 1978. ——*B. glaucescens* (Willd.) Sieb. ex Munro cv. Fernleaf (R. A. Young) Chia et But in Phytologia. 52(1): 258. 1982; But et al., Bamboos in Hongkong 38. 1985; Chia et al., Chinese bamboos 22. 1988.——*B. multiplex* (Lour.) Raeuschel ex J. A. et J. H. Schult. f. *fernleaf* (R. A. Young) Yi in Yi et al. Icon. Bamb. Sin. 129. 2008, et in Clav. Gen.Spec. Bamb.Sin. 40. 2009; Shi et al. in The Ornamental Bamb. in China. 291. 2012.

**特征**：与孝顺竹特征相似，主要区分在于其植株较小，秆高 3～6 m；小枝稍下弯，具叶 9～13，羽状排列，形似凤尾；叶片长 3.3～6.5 cm，宽 4～7 mm。

**用途**：著名观赏竹，供园林栽培或制作盆景。

分布：中国（华东、华南、西南以至台湾、香港均有栽培）。

5）篱笆竹 *Bambusa multiplex* 'Floribunda'

又名：Houou-chiku（日语）； Fernleaf Hedge Bamboo（English）

异名：*Barnbusa elegans*

*Bambusa glaucescens* f. *elegans*

*Bambusa multiplex* var. *elegans*

*Bambusa nana* var. *disticha*

*Leleba elegans*

引证：*Bambusa multiplex* 'Floribunda', Ohrnb., The Bamb. World. 269. 1999. ——*B. multiplex* var. *elegans* (Koidzumi) Muroi in Sugimoto, New Keys Jap. Tr., 1961:457. ——*Bambusa multiplex* 'Wang Tsai', S. Dransfield & E. A. Widjaja, Pl. Resources S. E. Asia, 7, 1995: 66. ——*B. elegans* Koidzumi ex Murata in Kitamura & Murata, Col. II1. Woody Pl. Jap., 2, 1979: 369.——*B. glaucescens* f. *elegans* (Koidzumi) Muroi & Sugimoto exMuroi & H. Okamura, Take sasa, 1977:147, 66.——*B. nana* var. *disticha* hort. ex R. A. Young in Nation. Hort. Mag. 25, 1946: 261.

特征：与凤尾竹特征相似，主要区分在于其叶不明显呈羽毛状排列，且植株相对矮小；叶片呈簇状，细小，多数挤向小枝的顶端。

用途：园艺栽培供观赏。

分布：日本（本州）；欧洲；美国；中国。

6）金色女神竹 *Bambusa multiplex* 'Golden Goddess'

异名：*Bambusa glaucescens* 'Golden Goddess'

*Bambusa glaucescens* var. *lutea*

*Bambusa multiplex* var. *lutea*

引证：*Bambusa multiplex* 'Golden Goddess', S. Dransfield & E. A. Widjaja, Pl. Resources S.E. Asia, 7, 1995:66; Ohrnb., The Bamb. World. 267. 1999; American Bamboo Society. *Bamboo Species Source List* No. 33: 8. Spring 2013. ——*B. glaucescens* 'Golden Goddess', Haubrich, 2, 1981. ——*B. multiplex* var. *lutea* Wen, 1982:31. ——*Bambusa glaucescens* var. *lutea* (Wen) Wen, 1985:16.

特征：与孝顺竹特征相似，主要区分在于其植株相对矮小，秆高3～

3.1 m，直径 1～1.3 cm；节间金黄色；叶较大。

**用途**：园艺栽培供观赏。

**分布**：欧洲；美国（佛罗里达州）；中国。

7）毛凤凰竹 *Bambusa multiplex* 'Incana'

**异名**：*Bambusa multiplex* var. *incana*

*Bambusa strigosa*

**引证**：*Bambusa multiplex* (Lour.) Raeuschel ex J. A. et J. H. Schult. var. *incana* B. M. Yang，Nat. Sci. Journ. Hunan Norm. Univ. 1983 (1): 77. f. 1. 1983; Keng et Wang in Flora Reip. Pop. Sin. 9(1): 110. 1996; Yi et al. Icon. Bamb. Sin. 127. 2008.——*Bambusa strigosa* Wen in Journ. Bamb. Res. 1(1): 31. pl. 8. 1982.

**特征**：与孝顺竹特征相似，主要区分在于其箨鞘背面被糙伏毛。

**用途**：园林栽培供观赏。

**分布**：中国（江西、湖南）。

8）金明孝顺竹 *Bambusa multiplex* 'Kimmei-Suhou'

**又名**：KJmmei-suhou（日本）

**异名**：*Bambusa glaucescens* f. *kimrnei-suhou*

*Barnbusa multiplex* f. *kimmei-suhou*

**引证**：*Bambusa multiplex* 'Kimmei-Suhou', Ohrnb., The Bamb. World. 268. 1999.——*B. glaucescens* f. *kimrnei-suhou* Muroi & Ka-sahara, 1972:7.——*Barnbusa multiplex* f. *kimmei-suhou* Muroi & Kasahara ex Muroi in J. Himeji Gakuin Wom. Coll. No. 1, 1974.

**特征**：与孝顺竹特征相似，主要区分在于其秆黄色，少数具绿色窄条纹。

**用途**：园林栽培供观赏。

**分布**：日本。

9）美岛竹 *Bambusa multiplex* 'Midori'

**又名**：Midori-hou-shiyou（日本）

**异名**：*Bambusa glaucescens* f. *midori*

*Barnbusa glaucescens* f. *alphonso-karrii* 'Midori'

*Bambusa glaucescens* 'Midori'

*Bambusa multiplex* f. *midori*

引证：*Bambusa multiplex* 'Midori', Ohrnb., The Bamb. World. 267. 1999.——*B. glaucescens* f. *midori* Muroi & Sugimoto, 10, 1971.——*B. multiplex* f. *midori* Muroi & Sugimoto ex Muroi in J. Himeji Gakuin Wom. Coll. no. 1, 1974: 2, as syn.——*B. glaucescens* 'Midori', Stover. 34.1983.——*Bambusa glaucescens* f. *alphonso-karrii* 'Midori'; Muroi & Sugimoto ex H. Okamura & M. Konishi in H. Okamura & Y. Tanaka, Hort. Bamb. Sp. Jap., 95.1986.

特征：与孝顺竹特征相似，主要区分在于其秆高 4～4.6 m，直径 3～3.8 cm；秆节间黄色并具宽窄不等的绿色纵条纹。

用途：园艺栽培供观赏。

分布：日本。

10）绿纹美岛竹 *Bambusa multiplex* 'Midori Green'

引证：*Bambusa multiplex* 'Midori Green', American Bamboo Society. *Bamboo Species Source List* No. 33: 8. Spring 2013.

特征：与美岛竹特征相似，主要区分在于其秆及分枝亮绿色，但具暗绿色纵条纹。

用途：园艺栽培供观赏。

分布：日本。

11）毛鞘银丝竹 *Bambusa multiplex* 'Pubivagina'

异名：*Bambusa multiplex* var. *pubivagina*

引证：*Bambusa multiplex* (Lour.) Raeuschel ex J. A. et J. H. Schult. var. *pubivagina* W. T. Lin et Z. J. Feng in Journ. Bamb. Res. 12(2): 33.1993; Yi et al. in Icon. Bamb. Sin. 127. 2008.

特征：与孝顺竹特征相似，主要区分在于其秆节间绿色，具白色纵条纹；箨鞘背面被很密白色或棕色交织的绒毛。

用途：园林栽培供观赏。

分布：中国（广东平远、五指石）。

12）观音竹 *Bambusa multiplex* 'Riviereorum'

又名：Riviere Hedge Bamboo（英语）

异名：*Bambusa glaucescens* var. *riviereorum*

*Bambusa multiplex* var. *nana*

*Bambusa multiplex* var. *riviereorum*

*Bambusa sciptoria*

引证：*Bambusa multiplex* 'Riviereorum', Ohrnb., The Bamb. World. 269. 1999; American Bamboo Society. *Bamboo Species Source List* No. 33: 9. Spring 2013.——*B. multiplex* (Lour.) Raeuschel ex J. A. et J. H. Schult. var. *riviereorum* R. Maire, Fl. Afr. Nord. 1: 355. 1952;Flora of China 22: 31. 2006; Yi et al. Icon. Bamb. Sin. 130. 2008, et in Clav. Gen.Spec. Bamb.Sin. 39. 2009; Shi et al. in The Ornamental Bamb. in China. 290. 2012.——*B. glaucescens* (Willd.) Sieb. ex Munro var. *riviereorum* (R. Maire) Chia et H. L. Fung in Phytologia 52(4): 257. 1982; But et al., Bamboos in Hongkong. 39. 1985.——*B. multiplex* (Lour.) Raeusch. var. *nana* auct. non (Roxb.) Keng f.: Flora of Guangzhou. 771. 1956.——*B. sciptoria* auct. non Dentist.: A. et C. Riv. in Bull. Soc. Acclim. III. 5: 685. 1878.

特征：与凤尾竹特征相似，主要区分在于其植株更加矮小，秆高1~3 m，直径3~5 mm，节间实心；小枝具叶13~23，弓状下弯；叶片长1.6~3.2 cm，宽2.5~6.5 mm。抗冻，可耐−8℃。

用途：著名观赏竹，适于栽培供观赏或作绿篱。

分布：中国（华中及西南各地）；印度尼西亚和泰国有栽培；欧洲；非洲。

13）石角竹 *Bambusa multiplex* 'Shimadai'

异名：*Bambusa glaucescens* var. *shimadai*

*Bambusa multiplex* var. *shimadai*

*Bambusa shimadai*

*Leleba shimadai*

引证：*Bambusa multiplex* (Lour.) Raeuschel ex J. A. et J. H. Schult. var. *shimadai* (Hayata) Sasaki in Trans. Nat. Hist. Soc, Form. 21: 118. 1931; Keng et Wang in Flora Reip. Pop. Sin. 9(1): 110. 1996; Yi et al. Icon. Bamb. Sin. 127. 2008. ——*Bambusa shimadai* Hayata, Icon. Pl. Form. 6: 151. f. 59. 1916. 竹内叔雄，竹的研究（中译本）110. 1957; W. C. Lin in Bull. Taiwan For. Res. Inst. No. 98; 24. ff. 15, 16. 1964. —— *Leleba shimadai* (Hayata) Nakai in Journ. Jap. Bot. 9(1): 17. 1933; W. C. Lin in 1. c. No.69: 65. ff. 30, 31. 1961. —— *Bambusa glaucescens* (Willd.) Sieb. ex Munro var. *shimadai* (Hayata) Chia et But in

Phytologia 52(4): 258. 1982.

**特征**：与孝顺竹特征相似，主要区分在于箨鞘先端近于两侧对称的宽拱形。

**用途**：园林栽培供观赏，亦常作绿篱和防风林。

**分布**：中国（广东，台湾）。

14）希罗竹 *Bambusa multiplex* 'Shirosuji'

**又名**：Shirosuji-kama，Shirosuji-bakama（日本）

**异名**：*Bambusa glaucescens* 'Shirosuji'

*Bambusa glaucescens* f. *shirosuji*

*Bambusa multiplex* f. *shirosuji*

**引证**：*Bambusa multiplex* 'Shirosuji', Ohrnb., The Bamb. World. 268. 1999. ——*B. multiplex* f. *shirosuji* Muroi & H. Okamura ex Muroi in J. Himeji Gakuin Wom. Coll. No. 1, 1974: 2. ——*B. glaucescens* 'Shirosuji', Stover, 1983:34. ——*B. glaucescens* f. *shirosuji* Muroi & H. Okamura, 1972:7.

**特征**：与孝顺竹特征相似，主要区分在于其秆节间具一细而长的白色纵条纹。

**用途**：园林栽培供观赏。

**分布**：日本。

15）什约克竹 *Bambusa multiplex* 'Shyokomachi'

**又名**：Shiyou-komachi（日本）

**异名**：*Bambusa glaucescens* f. *shyokomachi*

**引证**：*Bambusa multiplex* 'Shyokomachi', Ohrnb., The Bamb. World. 267. 1999. ——*B. glaucescens* f. *shyokomachi* Muroi & Maruyama ex Murci & H. Okamura, Take sasa, 1977: 149, 69.

**特征**：与孝顺竹特征相似，主要区分在于其叶蓝绿色。

**用途**：园林栽培供观赏。

**分布**：日本。

16）银丝竹 *Bambusa multiplex* 'Silverstripe'

**异名**：*Bambusa dolichomerithalla* 'Silverstripe'

*Bambusa floribunda* f. *albo-variegata*

*Bambusa glaucescens* cv. Silverstripe

*Bambusa multiplex* 'Albovariegata'

*Bambusa multiplex* f. *silverstripe*

*Bambusa multipex* var. *elegans* f. *albo-varegata*

*Bambusa multiplex* var. *silverstripe*

*Bambusa nana*. var. *albo-variegata*

*Leleba floribunda* f. *albo-variegata*

引证：*Bambusa multiplex* 'Silverstripe', American Bamboo Society. *Bamboo Species Source List* No. 33: 9. Spring 2013.——*B. multiplex* (Lour.) Raeusch. cv. Silverstripe R. A. Young in USDA Agr. Handb. No. 193: 41. 1961; Keng et Wang in Flora Reip. Pop. Sin. 9(1): 112. 1996. ——*B. dolichomerithalla* 'Silverstripe', Lin in Bull. Taiwan For. Res. Inst. No.271: 1976; ——*B. multiplex* 'Albovariegata', Ohrnb., The Bamb. World. 266. 1999. ——*B. multiplex* (Lour.) Raeuschel ex J. A. et J. H. Schult. f. *silverstripe* (R. A. Young) Yi in Yi et al. Icon. Bamb. Sin. 130. 2008, et in Clav. Gen.Spec. Bamb.Sin. 39. 2009; Shi et al. in The Ornamental Bamb. in China. 290. 2012. ——*B. nana* f. *albo-variegata* Makino in Journ. Jap. Bot. 1: 28. 1917. ——*B. nana* Roxb. var. *albo-variegata* E. G. Camus, Bambusa. 121. 1913. ——*B. floribunda* (Buse) Zoll. et Maur. ex Sieb. f. *albo-variegata* Nakai in Riko Kyoiku. 15: 66. 1932. ——*Leleba floribunda* (Buse)Nakai f. *albo-variegata* Nakai in Journ. Jap. Bot. 9: 12.1933.——*B. multiplex* var. *silverstripe* R. A. in Nat'l. Hort. Mag. 25: 260, 264. 1946.——*B. multipex* var. *elegans* (Koidz.) Muroi f. *albo-varegata* (Makino) Muroi ex Sugimoto, New Keys Jap. Trees. 457. 1961. ——*B. glaucescens* (Willd.) Sieb. ex Munro cv. Silverstripe (R. A. Young) Chia et But in Phytologia 52(4): 259. 1982; But et al., Bamboos in Hongkong 40, 1985.

特征：与孝顺竹特征相似，主要区分在于其秆下部节间、新鲜箨鞘和少数叶片均为绿色，并具白色纵条纹。

用途：同孝顺竹，但更具观赏性，适于庭园栽培观赏。

分布：中国（广东，香港）。

17）实心孝顺竹 *Bambusa multiplex* 'Solida'

又名： Komachi-dake, Houbi-chiku（日本）

**异名**：*Bambusa multiplex* f. *solida*

*Bambusa glaucescens* f. *solida*

**引证**：*Bambusa multiplex* 'Solida', Ohrnb., The Bamb. World. 269. 1999. ——*B. multiplex* f. *solida* Muroi & I. Maruyama in Sugimoto, 1961. ——*B. glaucescens* f. *solida* (Muroi & I. Maru-yama) Muroi & Sugimoto ex Muroi & H. Okamura, Take sasa, 149, 69. 1977.

**特征**：与孝顺竹特征相似，主要区分在于其秆高 3～5 m，直径 1～1.5 cm，秆节间实心或近实心；小枝多叶，叶长 1～9 mm，卷曲或仅尖端卷曲。

**用途**：园艺栽培供观赏。

**分布**：日本（中部，本州南部岛屿，西部日本海岸一部分，九州北部，四国岛南部）；中国（华中地区）。

18）小叶琴丝竹 *Bambusa multiplex* 'Stripestem Fernleaf'

**又名**：Beni-houou-chiku（日本）；Stripestem Femleaf Hedge Bamboo（英语）

**异名**：*Bambusa floribunda* f. *viridi-striata*

*Bambusa glaucescens* cv. Stripestem Fernleaf

*Barnbusa glaucescens* f. *viridistriata*

*Bambusa multiplex* f. *stripestem fernleaf*

*Bambusa multiplex* 'Stripestem'

*Bambusa multiplex* f. *viridistriata*

*Bambusa multiplex* var. *elagans* f. *viridi-striata*

*Bambusa multiplex* var. *stripestem fernleaf*

*Bambusa nana* f. *viridi-striata*

*Bambusa nana* var. *typica* f. *viridi-striata*

*Leleba floribunda* f. *viridi-striata*

**引证**：*Bambusa multiplex* 'Stripestem Fernleaf', Ohrnb., The Bamb. World. 267. 1999; American Bamboo Society. *Bamboo Species Source List* No. 33: 8. Spring 2013. ——*B. multiplex* (Lour.) Raeusch. cv. Stripestem Fernleaf R. A. Young in USDA Agr. Handb. No. 193: 41. 1961; Fl. Taiwan 5: 757. 1978. ——*B. multiplex* 'Stripestem', R. A. Young ex Lin in Bull. Taiwan For. Res. Inst. No. 271, 1976:44. ——*B. multiplex* var. *stripestem fernleaf* R. A. Young in Nation.

Hort. Mag. 25: 261. 1946. ——*B. multiplex* var. *elegans* (Koidz.) Muroi f. *viridi-striata* (Makino ex Tsuboi) Muroi ex Sugimoto, New Keys Jap. Trees 457. 1961; S. Suzuki, Index Jap. Bamb., 1978: 104, 340. ——*B. multiplex* (Lour.) Raeuschel ex J. A. et J. H. Schult. f. *stripestem fernleaf* (R. A. Young) Yi in Yi et al. Icon. Bamb. Sin. 130. 2008, et in Clav. Gen. Spec. Bamb.Sin. 39. 2009; Shi et al. in The Ornamental Bamb. in China. 290. 2012. ——*B. multiplex* f. *viridistriata* Muroi & Sugimoto ex Muroi in J. Himeji Gakuin Wom. Coll. No. 1, 1974. ——*Bambusa glaucescens* 'Stripestem Fernleaf', Hatusima, Woody Pl. Jap., 1976:316. ——*B. glaucescens* (Willd.) Sieb. ex Munro cv. Stripestem Fernleaf (R. A. Young) Chia et But in Phytologia 52(4): 259. 1982. ——*B. glaucescens* f. *viridistriata* (? Makino ex Tsuboi) Muroi & Sugimoto, 1971: 10. ——*B. nana* Roxb. var. *typica* Makino ex Tsuboi f. *viridi-striata* Makino ex Tsuboi, Ill. Jap. Sp. Bamb. ed. 2: 44. pl. 45. 1916. ——*B. nana* f. *viridi-striata* Makino in Journ. Jap. 1: 28. 1917. ——*B. floribunda* (Buse) Zoll. et Maur. ex Steud. f. *viridi-striata* Nakai in Riko Kyoiku 15; 66. 1932. ——*Leleba floribunda* (Buse) Nakai f. *viridi-striata* Nakai in Journ. Jap. Bot. 9: 12. 1933.

**特征：**与凤尾竹特征相似，主要区分在于其植株更加矮小，高 1~3 m，秆初时淡红色，后变为黄色并具宽度不等的绿色纵条纹；小枝下弯，具叶 12~20，叶片长 1.6~3.8 cm。

**用途：**珍稀观赏竹，适于盆栽或公园、小区、庭院栽培观赏。

**分布：**中国（香港，台湾）；日本；欧洲；美国。

19）小蕨竹 *Bambusa multiplex* 'Tiny Fern'

**又名：**Tiny Fern Bamboo（英语）

**异名：***Bambusa glaucescens* 'Tiny Fern'

**引证：***Bambusa multiplex* 'Tiny Fern', Ohrnb., The Bamb. World. 267. 1999; American Bamboo Society. *Bamboo Species Source List* No. 33: 9. Spring 2013. ——*B. glaucescens* 'Tiny Fern', Haubrich, 1981: 10.

**特征：**与凤尾竹特征相似，主要区分在于其植株特别矮小，秆高仅 0.6~0.9 m；叶片特别细小，其长多在 2.5 cm 以下。

**用途：**珍稀观赏竹，尤其适于盆栽观赏。

分布：美国。

20）银纹竹 *Bambusa multiplex* 'Variegata'

又名：Hou-shiyou-chiku，Taiho-chiku（日本）；Silverstripe Hedge Bamboo（英语）

异名：*Bambusa argentea* var. *vittata*
　　　*Bambusa glaucescens* 'Variegata'
　　　*Bambusa glaucescens* f. *variegata*
　　　*Barnbusa multiplex* f. *variegata*
　　　*Bambusa multiplex* f. *vittato-argentea*
　　　*Bambusa nana* var. *argenteostriata*
　　　*Bambusa nana* var. *normalis* f. *vittato-argentea*
　　　*Barnbusa nana* var. *normalis* f. *vittatoargentea*
　　　*Barnbusa nana* var. *vatiegata*
　　　*Bambusa scriptionis*
　　　*Bambusa vittato-argentea*
　　　*Leleba multiplex* f. *variegata*

引证：*Bambusa multiplex* 'Variegata', Ohrnb., The Bamb. World. 268. 1999.——*B. multiplex* f. *variegata* (Camus) R.A. Young ex A. V. Vasirev, 1956:29. ——*B. multiplex* f. *vittato-argentea* Nakai in Rika Kyo-iku 15 (6), 1932:67. ——*B. glaucescens* 'Variegata', Hatusima, Woody Pl. Jap., 1976:316. ——*B. glaucescens* f. *variegata* (Camus) Muroi & Sugimoto, 1971 : 10. ——*B. nana* var. *argenteostriata* hort. ex R.A. Young in Nation. Hort. Mag. 25, 1946: 260. ——*B. nana* var. *normalis* f. *vittato-argentea* Makino in S. Honda, Descr. Prod. For. Jap., 1900: 37. ——*B. nana* var. *normalis* f. *vittatoargentea* Makino ex Tsuboi, Illus. Jap. Sp. Bamb., 1916: 45, pl. XLVll. ——*B. nana* var. *vatiegata* Camus, Bamb., 1913: 12. ——*B. scriptionis* hort. ex W. Watson, 1889: 299. ——*B. vittato-argentea* hort. ex Mitford, Bamb. Gard., 1896: 55, 216.——*B. argentea* var. *vittata* Beadle; R. A. Young in Nation. Hort. Mag. 25, 1946: 260. ——*Leleba multiplex* f. *variegata* (Camus) Nakai in J. Jap. Bot. 9, 1933:16.

特征：与孝顺竹特征相似，主要区分在于其秆节间有窄奶油或白色纵条纹；箨鞘新鲜时亦具白色纵条纹；叶片具奶油色或白色条纹。

**用途**：园艺栽培供观赏。

**分布**：日本；美国；欧洲；澳大利亚。

21）垂柳竹 *Bambusa multiplex* 'Willowy'

**又名**：垂枝竹（中国）；Willowy Hedge Bamboo（英语）

**异名**：*Bambusa multiplex* f. *willowy*

**引证**：*Bambusa multiplex* 'Willowy', R. A. Young in Nation. Hort. Mag. 25, 1946: 260, 266; American Bamboo Society. *Bamboo Species Source List* No. 33: 9. Spring 2013; Ohrnb., The Bamb. World. 268. 1999; ——*B. multiplex* (Lour.) Raeusch. cv. Willowy R. A. Young, in USDA Agr. Handb. No.193: 42. 1961; Keng et Wang in Flora Reip. Pop. Sin. 9(1): 113. 1996. ——*B. multiplex* (Lour.) Raeuschel ex J. A. et J. H. Schult. f. *willowy* (R. A. Young) Yi in Yi et al. Icon. Bamb. Sin. 131. 2008, et in Clav. Gen.Spec. Bamb.Sin. 40. 2009; Shi et al. in The Ornamental Bamb. in China. 291. 2012.

**特征**：与孝顺竹特征相似，主要区分在于其分枝及叶下垂，叶片细长，一般长 10～20 cm，宽 8～16 mm。

**用途**：栽培供观赏，其叶片细长，枝叶下垂，形似垂柳，甚为美观。

**分布**：欧洲；中国（广东广州，四川宜宾）；美国亦有少量栽培。

22）黄条竹 *Bambusa multiplex* 'Yellowstripe'

**又名**：黄纹竹（中国）

**异名**：*Bambusa multiplex* f. *yellowstripe*

*Bambusa glaucescens* cv. Yellowstripe

**引证**：*Bambusa multiplex* 'Yellowstripe', Ohrnb., The Bamb. World. 267. 1999. ——*B. glaucescens* (Willd.) Sieb. ex Munro cv. Yellowstripe Chia et C. Y. Sia in Guihaia 8(1): 57. 1988; Keng et Wang in Flora Reip. Pop. Sin. 9(1): 111. 1996. ——*B. multiplex* (Lour.) Raeuschel ex J. A. et J. H. Schult. f. *yellowstripe* (Chia et C. Y. Sia) Yi in Yi et al. Icon. Bamb. Sin. 132. 2008，et in Clav. Gen. Spec. Bamb.Sin. 39. 2009.

**特征**：与孝顺竹特征相似，主要区分在于其秆节间在具芽或分枝一侧具黄色纵条纹。

**用途**：庭院栽培供观赏。

**分布**：中国（四川成都）。

### （6）撑篙竹 *Bambusa pervariabilis* McClure

秆高 7~10 m，直径 4~5.5 cm，梢端近直立；节间长约 30 cm，初时薄被白粉及有糙硬毛，基部数节间具黄绿色纵条纹；秆基部数节箨环上下各具一圈灰白色绢毛。秆每节具芽；分枝常自秆基部第一节开始，每节数枚至多枚簇生，3 枚主枝粗长。箨鞘早落，薄革质，背面无毛或有时被糙硬毛，新鲜时具黄绿色纵条纹，先端向外侧一边下斜而成不对称的拱形；箨耳不相等，波状皱褶，边缘具波曲状继毛，大耳沿箨鞘顶端向下倾斜，长 3.5~4 cm，宽约 1 cm，小耳近圆形或椭圆形，长约 1.5 cm，宽约 0.8 cm；箨舌高 3~4 mm，边缘不规则齿裂或条裂，具短流苏状毛；箨片直立，易脱落，狭卵形，幼时背面具黄绿色纵条纹，贴生棕色刺毛，基部圆形收窄后向两侧外延而与箨耳相连，其相连部分 3~7 mm，基部宽度约为箨鞘顶端宽的 2/3。小枝具叶（4）5（6）枚；叶鞘具短缘毛；叶耳倒卵形至倒卵状椭圆形，边缘具继毛；叶舌高约 0.5 mm；叶片线状披针形，长 10~15 cm，宽 1~1.5 cm，背面密被短柔毛，次脉 5~6 对，小横脉不清晰。假小穗以数枚簇生于花枝各节，线形，长 2~5 cm；小穗含小花 5~10 朵，基部托以具芽苞片 2 或 3 片；小穗轴节间长约 4 mm；颖仅 1 片，长圆形，长 6 mm，无毛，具 9 脉，先端急尖；外稃长圆状披针形，长 12~14 mm，无毛，具 13~15 脉，先端锐尖；内稃与外稃近等长或稍短，具 2 脊，脊向顶端被短纤毛，脊间 6 脉，脊外每边各 3 脉；鳞被 3，不相等，边缘被长纤毛，前方 2 片偏斜，长 2.7 mm，后方 1 片稍大，倒卵状长圆形，长 3 mm；花丝短，花药长 5 mm；子房长圆形，长约 1 mm，顶端被短硬毛，花柱长 1 mm，被短硬毛，柱头 3，长 3 mm，被毛。颖果幼时宽卵球状，长 1.5 mm，顶端被短硬毛，并有残留花柱和柱头。笋期 6~7 月。

村落及园林中常有栽培；秆供建筑、家具、农具等用材；秆表面刮制的"竹茹"供药用。

中国（广东，广西，福建，四川）。

#### 1）花身竹 *Bambusa pervariabilis* 'Multistriata'

**又名**：Punting Pole Bamboo（英语）

异名：*Bambusa pervariabilis* var. *multistriata*

引证：*Bambusa pervariabilis* 'Multistriata', Shi et al. in World Bamb. Ratt. 14(6): 27. 2016.——*B. pervariabilis* McClure var. *multistriata* W. T. Lin in Journ. Bamb. Res. 16 (3): 25. 1997; Yi et al. Icon. Bamb. Sin. 136. 2008, et in Clav. Gen. Spec. Bamb.Sin. 38. 2009; Shi et al. in The Ornamental Bamb. in China. 287. 2012.

特征：与撑篙竹特征相似，不同之处在于其秆和箨鞘有较多的白色纵条纹；节内和箨环下无一圈白色绢毛。

用途：同撑篙竹。

分布：中国（广西，广东广州，四川都江堰，香港）。

2）花撑篙竹 *Bambusa pervariabilis* 'Viridi-striata'

异名：*Bambusa pervariabilis* var. *viridi-striata*

*Bambusa pervariabilis* var. *viridistriata*

引证：*Bambusa pervariabilis* 'Viridi-striata', American Bamboo Society. *Bamboo Species Source List* No. 33: 9. Spring 2013.——*B. pervariabilis* McClure var. *viridi-striata* Q. H. Dai et X. C. Liu in Yi et al. Icon. Bamb. Sin. 136. 2008, et in Clav. Gen.Spec. Bamb.Sin. 38. 2009; Shi et al. in The Ornamental Bamb. in China. 287. 2012.——*B. pervariabilis* var. *viridistriata*, Ohrnb., The Bamb. World. 273. 1999.

特征：与撑篙竹特征相似，不同之处在于其秆黄色，具绿色纵条纹。

用途：栽培供观赏。

分布：中国（广西南宁、融水）。

**（7）硬头黄竹 *Bambusa rigida* Keng et Keng f.**

秆高 7~12 m，直径 3.5~6 cm；节间长 30~45 cm，初时薄被白粉，无毛，秆基部第一节箨环上方具一圈灰白色绢毛。分枝常自秆基部第 3、4 节开始，每节多枝簇生，主枝较粗长。箨鞘早落，背面无毛，有时下半部近内侧边缘处贴生暗棕色小刺毛，先端稍不对称的宽弧拱形；箨耳不等称，稍皱褶，边缘具波曲繸毛，大耳卵形，长约 2.5 cm，宽 1.5 cm，小耳卵形或近圆形，大小约为大耳的 2/3；箨舌高 2.5~3 mm，条裂，边缘具流苏状毛；箨片直立，易脱落，卵状三角形至卵状披针形，背面贴生极稀疏棕色小刺毛，腹面基部密生棕色小刺毛，基部圆形收窄后向两侧外延而与箨耳相连，其相连部分 3～4 mm，基部宽度约为箨鞘顶端宽的 2/5，边缘近基部被短纤毛。叶耳椭圆形，边缘具少数继毛；叶舌高约 0.5 mm；叶片长 7.5～18 cm，宽 1～2 cm，上面无毛或近基部被疏毛，下面密被短柔毛。假小穗淡绿色，单生或以数枚乃至多枚簇生于花枝各节，当多枚簇生成丛时其中多为不孕小穗，单生者则多为发育良好的孕性小穗，后者长 3～4.5 cm，含小花 3～7 朵，基部托以数枚具芽苞片；小穗轴节间形扁，无毛，长 2～4 mm，顶端膨大呈杯状；颖椭圆形，长 6～7 mm，多脉，先端急尖；外稃长圆状披针形，长 1～1.5 cm，宽 4～8 mm，具多脉，中脉隆起成脊，先端具短尖头；内稃较其外稃稍短，具 2 脊，脊于上部被纤毛，脊间 5 脉；鳞被 3，长 1.5～3 mm，上部边缘被长纤毛，前方 2 片半匙形，后方 1 片稍长，倒卵状披针形；花药长 4～6 mm，顶端被画笔状毛，子房具 3 棱，卵球形，具柄，连柄长 2～2.5 mm，顶部被糙硬毛，花柱被毛，长 1.5～2 mm，柱头 3，被短毛，长不及 1 mm。

秆材厚而坚实，且其下部挺直，常用作搭棚架、晒架以及做各种农具；生态营建，广泛栽培于河岸、丘陵、平坝及村庄附近。

**分布**：中国（四川，贵州北部，云南东北部和东南部，广东广州、福建厦门）。

1）黄条硬头黄竹 *Bambusa rigida* 'Luteolo-striata'

**异名**：*Bambusa rigida* f. *luteolo-striata*

**引证**：*Bambusa rigida* 'Luteolo-striata'，Shi et al. in World Bamb. Ratt. 14(6): 27. 2016.——*B. rigida* Keng et Keng f. f. *luteolo-striata* Yi et L. Yang in Journ. Sichuan For. Sci. Techn. 36(2):24. 2015.

**特征**：与硬头黄竹特征相似，不同之处在于其秆节间和箨鞘为绿色，均间有淡黄色纵条纹，枝条上有时亦具淡黄色条纹。

**用途**：同硬头黄竹，但更具观赏性。

**分布**：中国（四川长宁）。

**（8）青皮竹 *Bambusa textilis* McClure**

秆高 8～10 m，直径 3～5 cm；节间绿色，长 40～70 cm，幼时被白粉，贴生淡棕色刺毛，秆壁厚 2～3 mm。秆分枝较高，始于第 7 至第 11 节，各节多枝簇生，1 枚主枝稍粗长。箨鞘早落，稍有光泽，背面近基部贴生暗棕色刺毛，先端稍不对称的宽拱形；箨耳小，不相等，末端不外延，边缘具波曲细繸毛，大耳狭长圆形，稍向下倾斜，长约 1.5 cm，宽 4~5 mm，小耳长圆形，不倾斜，大小约为大耳的 1/2；箨舌高约 2 mm，边缘齿裂或条裂，被短纤毛；箨片直立，易脱落，卵状狭三角形，长度约为箨鞘的 2/3 或过之，背面近基部被暗棕色刺毛，腹面脉间具短刺毛或无毛而粗糙，基部稍心状收窄，其宽度约为箨鞘顶端的 2/3。叶耳发达，镰形，边缘具放射状弯曲繸毛；叶舌极低矮，边缘啮蚀状；叶片长 9~17 cm，宽 1~2 cm，下面密被短柔毛。假小穗单生或数枚乃至多枚簇生于花枝各节，鲜时暗紫色，干时古铜色，稍弯，线状披针形，长 3~4.5 cm，宽 5~8 mm；先出叶宽卵形，长 3 mm，具 2 脊，脊上无毛；具芽苞片 2 或 3 片，宽卵形，长 3~4.5 mm，无毛，先端急尖具短尖头；小穗含小花 5~8 朵，顶生小花不孕；小穗轴节间为半圆柱形或扁形，长约 4 mm，顶端膨大；颖仅 1 片，宽卵形，长 6 mm，无毛，具 21 脉，先端急尖具短尖头；外稃椭圆形，长 11~14 mm，无毛，具 25 脉，先端亦急尖具短尖头；内稃披针形，长 12~14 mm，常稍长于其外稃，具 2 脊，脊上无毛，脊间 10 脉，脊外每边各 4 脉；鳞被不相等，边缘被长纤毛，前方 2 片近匙形，长 3 mm，后方 1 片倒卵状椭圆形，长 2 mm；花丝细长，花药黄色，长 5 mm；子房宽卵球形，直径 2 mm，顶端增粗而被短硬毛，基部具子房柄，花柱长 0.7 mm，被短硬毛，柱头 3，长 6~7 mm，羽毛状。

1）青皮竹 Bambusa textilis 'Textilis'

**又名**：Wong chuk；Weaver's Bamboo（英语）

**异名**：*Bambusa textilis* var. *textilis*

**引证**：*Bambusa* 'Textilis', Shi et al. in World Bamb. Ratt. 14(6): 27. 2016. ——*B. textilis* McClure var. *textilis*, Keng et Wang in Flora Reip. Pop. Sin. 9(1): 124. 1996.

**特征**：与青皮竹种特征相似。

**用途**：秆为优良竹编用材，用于编制各种竹器和工艺品。中药"天竺黄"产自本竹的节间中。

**分布**：中国（西南、华东及华中地区）；美国。

2）矮青皮竹 *Bambusa textilis* 'Dwarf'

**引证**：*Bambusa textilis* 'Dwarf', American Bamboo Society. *Bamboo*

*Species Source List* No. 33: 10. Spring 2013.

**特征**：与青皮竹特征相似，不同之处在于其秆较矮小，高仅 5～5.5 m，直径 3～3.3 cm，且植株不甚挺拔。

**用途**：园艺栽培供编织或观赏。

**分布**：不详。

3）光秆青皮竹 *Bambusa textilis* 'Glabra'

**又名**：Smooth Weaver's Bamboo（英语）

**异名**：*Bambusa textilis* var. *glabra*

**引证**：*Bambusa textilis* 'Glabra', Shi et al. in World Bamb. Ratt. 14(6): 27. 2016.——*B. textilis* McClure var. *glabra* McClure in Lingnan. Univ. Sci. Bull. No. 9: 16. 1940; Keng et Wang in Flora Reip. Pop. Sin. 9(1):125.1996; Ohrnb., The Bamb. World. 276. 1999; Flora of China 22: 30. 2006; Yi et al. in Icon. Bamb. Sin. 143. 2008, et in Clav. Gen.Spec. Bamb.Sin. 44. 2009.

**特征**：与青皮竹特征相似，不同之处在于其秆节间幼时无毛；箨鞘背面无毛，箨片长度约为箨鞘长度的 1/2 或稍长于 1/2，且基部稍圆形收窄，箨舌高 1～1.5 mm。

**用途**：同青皮竹。

**分布**：中国（广东，广西，福建，香港）。

4）崖州竹 *Bambusa textilis* 'Gracilis'

**又名**：Slender Weaver's Bamboo（英语）

**异名**：*Bambusa textilis* var. *gracilis*

**引证**：*Bambusa textilis* 'Gracilis', Shi et al. in World Bamb. Ratt. 14(6): 27. 2016.——*B. textilis* McClure var. *gracilis* McClure in Lingnan. Univ. Sci. Bull. No. 9: 16. 1940; Keng et Wang in Flora Reip. Pop. Sin. 9(1):126. 1996; Ohrnb., The Bamb. World. 276. 1999; Flora of China 22: 30. 2006; Yi et al. in Icon. Bamb. Sin. 144. 2008, et in Clav. Gen.Spec. Bamb.Sin. 44. 2009.

**特征**：与青皮竹特征相似，不同之处在于其秆较细，直径常在 3 cm 以内；箨鞘背面近两侧及近基部处均疏生棕色刺毛，箨片长度约为箨鞘长的 1/2 或更短，其基部略圆形收窄，箨舌高约 1 mm。

**用途**：属优质观赏竹，丛型美观，姿态婀娜，适于公园、庭院、小区栽培观赏。

分布：中国（广东，广西，福建，四川）。

5）卡娜青皮竹 *Bambusa textilis* 'Kanapaha'

引证：*Bambusa textilis* 'Kanapaha', American Bamboo Society. *Bamboo Species Source List* No. 33: 10. Spring 2013.

特征：与青皮竹特征相似，不同之处在于其秆较高大，高 15～15.2 m，直径 6～6.4 cm，秆下半部光滑无枝且呈现蓝色。

用途：园艺栽培供观赏。

分布：美国。

6）紫斑竹 *Bambusa textilis* 'Maculata'

异名：*Bambusa textilis* f. *maculata*

*Bambusa textilis* var. *maculata*

引证：*Bambusa textilis* 'Maculata', Ohrnb. in The bamboos of the world. 276. 1999; American Bamboo Society. *Bamboo Species Source List* No. 33: 10. Spring 2013. —— *B. textilis* McClure cv. *maculata* Chia et al. in Guihaia 8(2): 127. 1988; Keng et Wang in Flora Reip. Pop.Sin.9(1):125.1996. ——*B. textilis* McClure f. *maculata* (McClure) Yi in Journ. Sichuan For. Sci. Techn. 28(3): 17. 2007; Yi et al. in Icon. Bamb. Sin. 144. 2008, et in Clav. Gen.Spec. Bamb.Sin. 43. 2009; Shi et al. in The Ornamental Bamb. in China 295. 2012.——*B. textilis* McClure var. *maculata* McClure in Lingnan Univ. Sci. Bull. No. 9: 16. 1940.

特征：与青皮竹特征相似，不同之处在于其秆基部数节间及箨鞘均具紫红色条状斑纹。

用途：较青皮竹色彩更丰富，适于公园、小区、庭院栽培观赏。

分布：中国（广西，广东广州）。

7）异变紫斑竹 *Bambusa textilis* 'Mutabilis'

引证：*Bambusa textilis* 'Mutabilis', American Bamboo Society. *Bamboo Species Source List* No. 33: 10. Spring 2013.

特征：与紫斑竹特征相似，不同之处在于其略显高大，通常秆高 12～12.2 m，直径 5～5.8 cm，秆节间特长。

用途：园艺栽培供观赏。

**分布**：不详。

8）紫秆竹 *Bambusa textilis* 'Purpurascens'

**异名**：*Bambusa textilis* f. *purpurascens*

*Bambusa textilis* var. *purpurascens*

**引证**：*Bambusa textilis* 'Maculata', Ohrnb. in The bamboos of the world. 276. 1999. —— *B. textilis* McClure cv. Purpurascens Chia et al. in Guihaia 8(2): 127. 1988; Keng et Wang in Flora Reip. Pop. Sin. 9(1):125.1996. ——*B. textilis* McClure f. *purpurascens* (N. H. Xia) Yi in Journ. Sichuan For. Sci. Techn. 28(3): 17. 2007; Yi et al. in Icon. Bamb. Sin. 144. 2008, et in Clav. Gen.Spec. Bamb.Sin. 43. 2009;Shi et al. in The Ornamental Bamb. in China. 296. 2012. ——*B. textilis* McClure var. *purpureascens* N. H. Xia in Bamb. Res. 1985(1): 38. f. 2. 1985.

**特征**：与青皮竹特征相似，不同之处在于其秆节间为绿色，间有宽窄不等的紫红色纵条纹且有时整枝竹秆呈紫红色。

**用途**：较青皮竹色彩更美丽，可与紫竹媲美，属于比较罕见的高品质观赏竹种，适于公园、小区、庭院栽培观赏。

**分布**：中国（广东，福建，四川，云南）。

9）斯克兰顿竹 *Bambusa textilis* 'Scranton'

**引证**：*Bambusa textilis* 'Mutabilis', American Bamboo Society. *Bamboo Species Source List* No. 33: 10. Spring 2013.

**特征**：与青皮竹特征相似，不同之处在于竹秆比较散开，秆高 9～9.1 m，直径 5～5.1 cm；秆分枝较短。

**用途**：园艺栽培供观赏。

**分布**：不详。

10）花青皮竹 *Bambusa textilis* 'Viridistriata'

**异名**：*Bambusa textilis* f. *viridi-striata*

**引证**：*Bambusa textilis* 'Viridistriata', Shi et al. in World Bamb. Ratt. 14(6): 27. 2016.——*B. textilis* McClure f. *viridi-striata* Yi in Journ. Bamb. Res. 21(1): 10. 2002; Yi et al. in Icon. Bamb. Sin. 146. 2008, et in Clav. Gen.Spec. Bamb.Sin. 43. 2009;Shi et al. in The Ornamental Bamb. in China. 295. 2012.

**特征**：与青皮竹特征相似，不同之处在于其秆节间黄色，具绿色纵条纹；箨鞘新鲜时为绿色且具黄色纵条纹。

**用途**：竹丛挺拔，色彩美丽，是近年新发现的非常美丽的高品质观赏竹种。

**分布**：中国（广西，四川）。

（9）马甲竹 *Bambusa tulda* Roxb.

秆高 8～10 m，直径 5～7 cm；节间长 40～45 cm，幼时被白粉，秆基部数节箨环上方具一圈灰白色绢毛，并生有气根。秆基部第一节开始分枝，每节多枝簇生，主枝粗长。箨鞘早落，厚革质，背面初时被白粉，密被暗褐色贴生刺毛，先端呈不对称的弧形，边缘具极短的纤毛；箨耳显著不相等，波状强皱褶，大耳明显下延至箨鞘全长的 1/3，长肾形或倒卵状披针形，长 4.5～5 cm，宽约 1.5 cm；箨舌高 1.5～2 mm，全缘；箨片直立，宽卵形，基部心形或圆形收窄并外展与箨耳相连，其相连部分 1～1.3 cm，腹面被糙硬毛或粗糙，基部宽度约为箨鞘顶端的 5/8，边缘近基部波状，具短缘毛。叶耳不发达或不存在，鞘口繸毛 1、2 或无；叶舌边缘微齿裂；叶片长 15～20 cm，宽 1.5～2.5 cm，上面有时基部被短硬毛，下面淡绿色，密被短柔毛。假小穗单生或以 2～5 枚簇生于花枝各节；小穗线形至线状披针形，长 2.5～7.5 cm，宽约 5 mm，含小花 4～6 朵，顶端有 1 或 2 朵不孕小花；小穗轴节间呈棒状，近内稃一面扁平，具条纹，顶端边缘被纤毛；颖 1 或 2 片，多脉，先端急尖；外稃卵形至长圆形，长 1.2～2.5 cm，宽 7.5 mm，无毛，亦具多脉，先端急尖或渐尖而具细短尖头，有时边缘稍被纤毛；内稃稍短于外稃，具 2 脊，脊上被纤毛，顶端具画笔状毛，脊间 5～7 脉；鳞被 3，长约 3.8 mm，前方 2 片基部增厚，具 5 脉，边缘被长纤毛，后方一片的基部不增厚；花药紫红色，长 7.5～10 mm，先端钝或微凹缺；子房倒卵形或倒卵状长圆形，顶部增厚而被长硬毛，花柱极短而被长硬毛，柱头 3，羽毛状。颖果长圆形，长 7.5 mm，腹面具纵沟槽状种脐，顶端被长硬毛。产中国广东、广西、西藏南部；福建、四川有引栽。

1）条纹马甲竹 *Bambusa tulda* 'Striata'

**引证**：*Bambusa tulda* 'Striata', American Bamboo Society. *Bamboo Species Source List* No. 33: 10. Spring 2013.

**特征**：与马甲竹特征相似，不同之处在于其秆高约 21.3 m，直径约

10.2 cm；秆具黄色纵条纹，且越是基部条纹越多。

**用途**：园艺栽培供观赏。

**分布**：不详。

### （10）青秆竹 *Bambusa tuldoides* Munro

秆高 6～10 m，直径 3～5 cm；节间长 30～36 cm，初时薄被白粉，无毛；秆基部第一至第二节箨环上下各具一圈灰白色绢毛。分枝常自秆基部第一节或第二节开始，主枝粗长。箨鞘早落，背面无毛，新鲜时靠外侧一边常具 1～3 条黄白色纵条纹，先端为不对称的宽拱形；箨耳不等大，大耳卵形或卵状椭圆形，长约 2.5 cm，宽 1～1.4 cm，稍皱褶，边缘具细弱繸毛，小耳卵圆形或椭圆形，大小约为大耳的一半；箨舌高 3～4 mm，条裂，边缘具短繸毛；箨片直立，易脱落，不对称卵状三角形至狭三角形，幼时两面均具棕色小刺毛，基部稍圆形收窄后向两侧外延而与箨耳相连，基部宽约为箨鞘顶端的 2/3～3/4，边缘近基部稍皱褶，具波曲状繸毛。叶耳缺失或存在；叶舌低矮，截平；叶片长 10～18 cm，宽 1.5～2 cm，上面无毛或近基部疏生柔毛，下面密被短柔毛，基部近圆形或宽楔形。假小穗以数枚簇生于花枝各节，簇丛基部托以鞘状苞片，淡绿色，稍扁，线状披针形，长 2～3 cm，宽 3～4 mm；先出叶具 2 脊，脊上被纤毛；具芽苞片 2 片，无毛，先端钝；小穗含小花 6 或 7 朵，位于上下两端者不孕，中间的小花为两性；小穗轴节间扁平，长 3～4 mm，顶端膨大呈杯状而被微毛；颖常 1 片，卵状长圆形，长 8.5 mm，无毛，先端急尖；外稃卵状长圆形，长 11～14 mm，具 19 脉，无毛，先端钝并具短尖头；内稃与其外稃近等长或稍短，两脊的上部疏生极短白色纤毛，近顶端的毛较长，脊间和脊外的每边均具 4 脉，并生有小横脉，先端略钝，并有一簇画笔状白毛；鳞被 3，倒卵形，边缘被长纤毛，前方 2 片偏斜，宽短，长 2.5 mm，后方一片狭长，长约 3 mm；花药长 3 mm，先端微凹；子房倒卵形，具柄，长 1.2 mm，顶部增厚并被长硬毛，花柱长 0.7 mm，被长硬毛，柱头 3，长 5.5 mm，羽毛状。颖果圆柱形，稍弯，长 8 mm，直径 1.5 mm，顶端钝圆而增厚，并被长硬毛和残留的花柱。笋期 7～9 月。

1）青秆竹 *Bambusa* 'Tuldoides'

**又名**：水竹，硬生桃竹，硬散桃竹，硬头黄竹（中国）；Buloh balai（马来语）；Bambu blenduk（印度尼西亚）；Verdant Bamboo，Punting Pole Bamboo

（英语）

**引证**：*Bambusa* 'Tuldoides', Shi et al. in World Bamb. Ratt. 14(6): 27. 2016.——*B. tuldoides* Munro cv. Tuldoides, Keng et Wang in Flora Reip. Pop. Sin. 9(1): 87. 1996.

**特征**：与青秆竹种特征相似。

**用途**：秆供建筑、家具、农具等用材。秆表面刮制的"竹茹"供药用。

**分布**：中国（广东，香港，广西，贵州南部，福建，云南，台湾）；越南；东南亚广泛栽培；南美洲亦有引栽。

2）鼓节竹 *Bambusa tuldoides* 'Swolleninternode'

**又名**：鼓节青秆竹（中国）；Butto-chiku（日本）；Buddha Bamboo（英语）

**异名**：*Bambusa tuldoides* f. *swolleninternode*
　　　　*Bambusa tuldoides* 'Ventricosa'

**引证**：*Bambusa tuldoides* 'Swolleninternode', Xia in Bamb. Res. No. 23, 1985: 38, fig.1; Shi et al. in World Bamb. Ratt. 14(6): 28. 2016.——*B. tuldoides* Munro cv. Swolleninternode N. H. Xia in Bamb. Res. 1985 (1): 38. f. 1. 1985; Keng et Wang in Flora Reip. Pop. Sin. 9(1):88.1996.——*B. tuldoides* Munro f. *swolleninternode* (N. H. Xia) Yi in Journ. Sichuan For. Sci. Techn. 28(3): 17. 2007; Yi et al. in Icon. Bamb. Sin. 147. 2008, et in Clav. Gen. Spec. Bamb. Sin. 38. 2009; Shi et al. in The Ornamental Bamb. in China. 288. 2012. ——*B. tuldoides* 'Ventricosa', Ohrnb., The Bamb. World. 277. 1999.

**特征**：与青秆竹特征相似，不同之处在于其秆下部节间短缩，并在基部肿胀呈鼓节状。

**用途**：竹丛紧凑，竹秆畸变，属观赏竹中之上品，适于盆栽或公园、庭院栽培观赏。

**分布**：中国（广东，福建）；东南亚地区；美国；欧洲。

3）金明鼓节竹 *Bambusa tuldoides* 'Ventricosa Kimmer'

**又名**：Kimmei-daifuku-chiku, Kimmei-butto-chiku（日本）

**异名**：*Bambusa tuldoides* f. *kimmei*

**引证**：*Bambusa tuldoides* 'Ventricosa Kimmei', Ohrnb., The Bamb. World. 278. 1999.

**特征**：与鼓节竹特征相似，不同之处在于其秆黄色，但具少数绿色条纹，秆分枝一侧沟槽浅绿色；叶具少量白色条纹。

**用途：** 竹丛紧凑，竹秆畸变，属观赏竹中之上品，适于盆栽或公园、庭院栽培观赏。

**分布：** 日本。

### （11）佛肚竹 *Bambusa ventricosa* McClure

正常秆高达 10 m，直径 5 cm，梢尾部略下垂，下部稍"之"字形曲折；节间长 30~35 cm，幼时无白粉，平滑无毛，秆下部箨环上下具灰白色绢毛环，秆壁厚 6~12 mm；箨环无毛。秆基部第三、四节上开始分枝，常 1~3 枝，其上小枝有时短缩为软刺，中上部各节具多枝，主枝 3，粗长。畸形秆高达 5 m，直径稍细，节间短缩并在下部肿胀呈瓶状，长达 6 cm；常具单枝，其节间明显肿胀。箨鞘早落，干时纵肋隆起，背面无毛，先端近于对称的宽拱形或截形；箨耳不相等，边缘具继毛，大耳宽 5~6 mm，小耳宽 3~5 mm；箨舌高 0.5~1 mm，边缘具很短流苏状毛；箨片直立，卵形或卵状披针形，基部稍心形收窄，宽度稍窄于箨鞘顶端。小枝具叶（5）7~11；叶耳卵形或镰形，边缘具继毛；叶舌近截形；叶片长 9~18 cm，宽 1~2 cm，下面密被短柔毛。假小穗单生或数枚簇生于花枝各节，线状披针形，稍扁，长 3~4 cm；先出叶宽卵形，长 2.5~3 mm，具 2 脊，脊上被短纤毛，先端钝；具芽苞片 1 或 2 片，狭卵形，长 4~5 mm，具 13~15 脉，先端急尖；小穗含两性小花 6~8 朵，其中基部 1 或 2 朵和顶生 2 或 3 朵小花常为不孕性；小穗轴节间形扁，长 2~3 mm，顶端膨大呈杯状，其边缘被短纤毛；颖常无或仅 1 片，卵状椭圆形，长 6.5~8 mm，具 15~17 脉，先端急尖；外稃无毛，卵状椭圆形，长 9~11 mm，具 19~21 脉，脉间具小横脉，先端急尖；内稃与外稃近等长，具 2 脊，脊近顶端处被短纤毛，脊间与脊外两侧均各具 4 脉，先端渐尖，顶端具一小簇白色柔毛；鳞被 3，长约 2 mm，边缘上部被长纤毛，前方两片形状稍不对称，后方 1 片宽椭圆形；花丝细长，花药黄色，长 6 mm，先端钝；子房具柄，宽卵形，长 1~1.2 mm，顶端增厚而被毛，花柱极短，被毛，柱头 3，长约 6 mm，羽毛状。

1）佛肚竹 *Bambusa ventricosa* 'Nana'

**又名：** 小佛肚竹、佛竹、葫芦竹

**异名：** *Bambusa tuldoides* 'Nana'

**引证**：*Bambusa ventricosa* McClure cv. Nana Wen in Journ. Bamb. Res. 4(2): 18. 1985; Keng et Wang in Flora Reip. Pop. Sin. 9(1): 70. 1996.——*B. tuldoides* 'Nana', Ohrnb., The bamboos of the world. 278. 1999.

**特征**：与佛肚竹种特征相似，不同之处在于其秆畸形比例较大，大多数节间均短缩呈佛肚状。

**用途**：节间短缩，形态奇异，是一种广泛栽培的观赏竹种。

**分布**：中国（南方各地普遍盆栽或地栽）；日本；泰国；越南等。

2）金明佛肚竹 *Bambusa ventricosa* 'Kimmei'

**又名**：金丝葫芦竹（中国）；Kimmei-daifuku-chiku, Kimmei-butto-chiku（日本）

**异名**：*Bambusa ventricosa* f. *kimmei*

**引证**：*Bambusa ventncosa* 'Kimmei', Muroi & Y. Tanaka ex H. Okamura in H. Okamura & Tanaka, 1986: 7. ——*B. ventricosa* McClure f. *kimmei* Muroi et Y. Tanaka ex H. Okamura & Y. Tanaka in Hort. Bamb. Sp. Jap., 7, 101, invalid (Engl. descr.). 1986; Yi et al. in Icon. Bamb. Sin. 112. 2008, Yi et al.in Clav. Gen.Spec. Bamb.Sin. 32. 2009; Shi et al. in The Ornamental Bamb. in China. 279. 2012.

**特征**：与畸形秆佛肚竹近似，不同之处在于其秆黄色，节间分枝一侧及周围具少数绿色纵条纹；部分叶片具淡黄白色纵条纹。

**用途**：节间膨大，竹型独特，色彩美丽，玲珑可人，是观赏竹中之上品，适于盆栽或制作竹盆景。

**分布**：中国（江苏南京，四川成都）；日本。

**（12）龙头竹 *Bambusa vulgaris* Schrader ex Wendland**

秆高 8～15 m，直径 5～9 cm，下部径直或稍"之"字形曲折；节间长 20～30 cm，初时稍被白粉，贴生淡棕色刺毛；秆基部数节节内具短气生根，并在箨环上下具灰白色绢毛。秆下部开始分枝，每节多枝簇生，主枝较粗长。箨鞘早落，背面密被棕黑色刺毛，先端弧拱形，但在与箨耳连接处弧形下凹；箨耳发达，近等大，长圆形或肾形，斜升，宽 8～10 mm，边缘具弯曲继毛；箨舌高 3～4 mm，边缘细齿裂，具极短细缘毛；箨片直立或外展，易脱落，宽三角形或三角形，背面疏被棕色小刺毛，腹面密被棕色小刺毛，基部稍圆形收窄，其宽度约为箨鞘顶端的 1/2，近基部边缘具弯曲细继毛。

叶鞘初时疏被棕色糙硬毛；叶耳如存在时常为宽镰形，边缘继毛少数或缺失；叶舌全缘；叶片长10～30 cm，宽1.3～2.5 cm，无毛，基部近圆形，稍不对称。假小穗数枚簇生于花枝各节；小穗稍扁，狭披针形至线状披针形，长2～3.5 cm，宽4～5 mm，含小花5～10朵，基部托以数片具芽苞片；小穗轴节间长1.5～3 mm；颖1或2片，背面仅于近顶端被短毛，先端具硬尖头；外稃长8～10 mm，背面近顶端被短毛，先端具硬尖头；内稃略短于外稃，具2脊，脊上被短纤毛；鳞被3，长2～2.5 mm，边缘被长纤毛；花药长6 mm，顶端具一小簇短毛；花柱细长，长3～7 mm，柱头短，3枚。

1）龙头竹 *Bambusa* 'Vulgaris'

**又名**：泰山竹（中国）；Bambu ampel（印度尼西亚）；Buloh aur，Buloh pau，Buloh minyak，Aur beting（马来语）；Mai-luang，Phai-luang（泰国）；Daisan-chiku（日本）；Gemeiner Bambus（德国）；Common Bamboo（英语）

**引证**：*Bambusa* 'Vulgaris', Shi et al. in World Bamb. Ratt. 14(6): 28. 2016. ——*Bambusa vulgaris* Schrader ex Wendland cv. Vulgaris, Keng et Wang in Flora Reip. Pop. Sin. 9(1): 96. 1996.

**特征**：与龙头竹种特征相似。

**用途**：秆作建筑、造纸或农业等用材。

**分布**：中国（云南，广西，广东，香港，福建）；亚洲热带地区；非洲马达加斯加岛。

2）黄金间碧竹 *Bambusa vulgaris* 'Vittata'

**又名**：青丝金竹，龙头竹（中国）；Buloh gading，Aur gading，Buloh kuning（马来语）；Bambu kuning（印度尼西亚）；KJnshi-chiku（日本）；Golden Common Bamboo（英语）

**异名**：*Bambusa striata*

*Bambusa vulgaris* f. *vittata*

*Bambusa vulgads* 'Striata'

*Bambusa vulgaris* var. *striata*

*Bambusa vulgaris* var. *vittata*

**引证**：*Bambusa vulgaris* 'Vittata', Hatusima, Woody Pl. Jap., 1976: 315; Ohrnb. in The bamboos of the world. 279. 1999.; American Bamboo Society. *Bamboo Species*

*Source List* No. 33: 11. Spring 2013.——B. vulgaris Schrader ex Wendland cv. Vittata (A. et C. Riv.) McClure in Agr. Handb. USDA. No. 193: 46. 1961; Keng et Wang in Flora Reip. Pop. Sin.9(1):97.1996.——B. striata Loddiges ex Lindley in Penny Cycl., 3, 1835:357.——B. vulgaris 'Vittata', McClure ap. Swallen in Fieldiana Bot. 24 (2), 1955:60. ——B. vulgaris var. *striata* (Loddiges ex Lindley) Gamble in Ann. Roy. Bot. Gard. Calcutta 7, 1896: 44, pl. 40 fig. 4-5. ——B. vulgaris Schrader ex Wendland f. *vittata* A. et C. Riv. 1982: 467.——B. vulgaris Schrader ex Wendland f. *vittata* (A. et C. Riv.) Yi in Journ. Sichuan For. Sci. Techn. 28(3): 17. 2007; Yi et al. in Icon. Bamb. Sin. 149. 2008, et in Clav. Gen.Spec. Bamb. Sin. 36. 2009; Shi et al. in The Ornamental Bamb. in China. 283. 2012.——B. vulgaris Schrader ex Wendland var. *vittata* A. et C. Riv. in Bull. Soc. Acclim. Ⅲ 5: 640. 1878.

**特征**：与龙头竹特征相似，不同之处在于其秆黄色，具绿色纵条纹，箨鞘新鲜时绿色，具黄色纵条纹。

**用途**：竹丛高大，竹型美观，色彩艳丽，属于著名大型丛生观赏竹，深受人们的喜爱，尤其适于公园、小区、风景区栽培观赏。

**分布**：中国（福建，台湾，广东，广西，香港，海南，云南）；世界各地（包括东亚、南亚、东南亚、马达加斯加等热带和亚热带地区）广泛栽培。

3）大佛肚竹 *Bambusa vulgaris* 'Wamin'

**又名**：Bambu blenduk（印度尼西亚）；Wamin Bamboo（英语）

**异名**：*Bambusa vulgaris* cv. Waminii

*Bambusa vulgaris* f. *wamin*

*Bambusa wamin*

**引证**：*Bambusa vulgaris* 'Wamin', Ohrnb. in The bamboos of the world. 280. 1999; American Bamboo Society. *Bamboo Species Source List* No. 33: 11. Spring 2013.——B. vulgaris Schrader ex Wendland cv. Wamin McClure, Keng et Wang in Flora Reip. Pop. Sin. 9(1): 97. 1996.——B. vulgaris Schrader ex Wendland f. *waminii* Wen in Journ. Bamb. Res. 4(2): 16. 1985; Yi et al. in Icon. Bamb. Sin. 150. 2008, et in Clav. Gen.Spec. Bamb.Sin. 35. 2009; Shi et al. in The Ornamental Bamb. in China. 282. 2012. ——B. wamin Brandis ex Camus, Bamb., 1913: 135.

**特征**：与龙头竹特征相似，不同之处在于其秆绿色或有时为淡黄绿色，下部各节间极度短缩，并在各节间基部大幅肿胀呈佛肚状。

**用途**：竹丛紧凑，竹秆畸变，形态独特，属于著名丛生异型观赏竹，是观赏竹中之上品，深受人们的喜爱，园林中常见栽培，尤其适于盆栽或制作竹盆景。

**分布**：中国（华南以及浙江、福建、台湾、四川西南部、云南南部等）；东南亚；美国；欧洲。

4）条纹大佛肚竹 *Bambusa vulgaris* 'Wamin Striata'

**引证**：*Bambusa vulgaris* 'Wamin Striata', American Bamboo Society. *Bamboo Species Source List* No. 33: 11. Spring 2013.

**特征**：与大佛肚竹特征相似，不同之处在于其秆亮绿色具深绿色纵条纹。

**用途**：园艺栽培供观赏。

**分布**：不详。

## 2.2 方竹属 *Chimonobambusa* Makino

### （1）狭叶方竹 *Chimonobambusa angustifolia* C. D. Chu & C. S. Chao

秆高 2～5 m，直径 1～2 cm；节间长 10～15 cm，圆筒形，或下部节间略呈四方形，幼时密被白色柔毛和稀疏刺毛，毛脱落后留有瘤基而略粗糙；箨环初时被淡褐色纤毛；秆环稍平坦或在分枝节上甚隆起；秆下部各节内具发达的气生根刺 9～14 条。秆每节上枝条 3 枚，但亦有多枝者，枝条实心，其节处强烈隆起。箨鞘短于节间，纸质至厚纸质，黄褐色，背面上部具大小不等的灰白色或淡黄色圆斑，下部具稀疏淡黄色柔毛及小刺毛，鞘缘密生黄褐色纤毛，纵肋明显，小横脉紫色，具缘毛；箨舌截形或拱形，甚至强烈拱状上突，全缘，有微小纤毛；箨片极小，锥状三角形，长 3～5 mm，箨耳缺或不发达。小枝具叶 1～3（4）；叶耳无或不发达，鞘口𫄧毛少数，仅 3～5 条，直立，长 3～5 mm；叶舌低矮；叶片线状披针形或线形，长 6～15 cm，宽 0.5～1.2 cm，次脉 3～4 对。花未见。笋期 8～9 月。

1）实心狭叶方竹 *Chimonobambusa angustifolia* 'Repleta'

**异名**：*Chimonobambusa angustifolia* f. *repleta*

**引证**：*Chimonobambusa angustifolia* C. D. Chu et C. S. Chao f. *repleta* Yi et H. R. Qi in Journ. Bamb. Res. 23(3): 6. 2004; Yi et al. in Icon. Bamb. Sin. 266.

2008, et in Clav. Gen.Spec. Bamb.Sin.79. 2009.

**特征：** 与狭叶方竹特征相似，不同之处在于其秆更矮小，高 1.8 m，直径 0.5 cm；秆及枝节间全为实心。

**用途：** 适于地栽或盆栽观赏。

**分布：** 中国（重庆梁平）。

**（2）寒竹 *Chimonobambusa marmorea*( Mitford ) Makino**

地下茎复轴。秆散生兼小丛生，高 1~1.5（3）m，直径 0.5~1 cm；节间圆筒形或在分枝一侧扁平并具沟槽，长 10~14 cm；箨环初时具棕褐色绒毛；秆环稍隆起；基部数节内具少数气生根刺。秆每节上枝条 3 枚。箨鞘宿存，稍长于节间，背面黄褐色，具灰白色斑块，无毛，或仅疏被淡黄色小刺毛，具微缘毛；箨耳无；箨舌截形或稍拱形；箨片锥状，直立，长 2~3 mm。小枝具叶 2、3；叶鞘鞘口䍁毛长 3~4 mm；叶片长 10~14 cm，宽 0.7~0.9 cm，无毛，次脉 4、5 对。花枝基部具一组逐渐增大的苞片，中、上部具假小穗 1~4 枚；假小穗线形，长 2~4 cm，苞片 0~2，腋内有芽或无芽；小穗含 4~7 花；小穗轴节间长 3~4 mm，无毛；颖片 1、2，或缺，长 6~8 mm，具 5~7 脉；外稃绿色或略带紫色，长 6~7 mm，具 5~7 脉及小横脉；内稃等长于外稃，背部具 2 脊，脊间具 2 脉，先端截平或微具 2 齿裂，脊外两侧各具 2 脉；鳞被 3；雄蕊 3，花丝分离；子房细长卵形，花柱 1，短，柱头 2，羽毛状。颖果果皮厚肉质，干后坚果状。

竹姿高雅，极为优美，为著名的庭园观赏竹种，尤其适宜盆栽。

中国（浙江、福建、四川、云南、广西）；日本；欧、美各国广泛引栽。

1）银明寒竹 *Chimonobambusa marmorea* 'Gimmei'

**又名：** Gimmei-kan-chiku（日本）

**异名：** *Chimonobambusa marmorea* f. *gimmei*

**引证：** *Chimonobambusa marmorea* 'Gimmei', Ohrnberger, Bamb. World. Chimonobambusa ed. 2, 1996:18; Ohrnb., The Bamb. World. 181. 1999.——*Ch. marmorea* (Mitford) Makino f. *gimmei* Muroi et Kasahara in Rep. Fuji Bamb. Gard. No. 17: 8, 1972; Muroi in J. Himeji Gakuin Wom. Coll. No. 1,1974: 2; H. Okamura & al., Il1. Hort. Bamb. Sp. Jap., 1991: 349; Yi et al. in Icon. Bamb. Sin. 272. 2008, et in Clav. Gen. Spec. Bamb.Sin.76. 2009.

**特征**：秆散生兼小丛生，高 1～1.5（3）m，直径 0.5～1 cm；秆绿色，节间具芽或分枝一侧沟槽具淡黄绿色纵条纹；节间圆筒形，长 10～14 cm；箨环初时具棕褐色绒毛；秆环稍隆起；基部数节内具少数气生根刺。秆每节上枝条 3 枚。箨鞘宿存，稍长于节间，背面黄褐色，具灰白色斑块，无毛，或仅疏被淡黄色小刺毛，具微缘毛；箨耳无；箨舌截形或稍拱形；箨片锥状，直立，长 2～3 mm。小枝具叶 2、3；叶片具白色纵条纹；叶鞘鞘口继毛长 3～4 mm；叶片长 10～14 cm，宽 0.7～0.9 cm，无毛，次脉 4、5 对。颖果果皮厚肉质，干后坚果状。

**用途**：竹姿高雅，极为优美，为著名的庭园观赏竹种，尤其适宜用作盆栽、盆景。

**分布**：日本；中国；欧、美各国有引栽。

2）金明寒竹 *Chimonobambusa marmorea* 'Kimmei'

**又名**：Suisho-chiku, Somoku-kinyoshyu（日本）

**异名**：*Chimonobambusa marmorea* f. *kimrnei*

**引证**：*Chimonobambusa marmorea* 'Kimmei', Ohrnberger, Bamb. World Gen. Chimonobambusa, 1990:26; Ohrnb., The Bamb. World. 182. 1999.——*Ch. marmorea* f. *kimmei* Muroi & H. Okamura in J. Himeji Gakuin Wom. Coll. No. 1, 1974: 2.

**特征**：与银明寒竹特征相似，不同之处在于其秆上条纹为金黄色。

**用途**：园艺栽培供观赏。

**分布**：日本。

3）花叶寒竹 *Chimonobambusa marmorea* 'Variegata'

**又名**：Chigo-kan-chiku（日本）；meaning small winter bamboo, Chiryo-kan-chiku, Beni-kan-chiku, Heisaku-kan-chiku（英语）

**异名**：*Arundinaria marmorea* 'Variegata'

*Arundinaria marmorea* var. *variegata*

*Chimonobambusa marrnorea* f. *albovatiegata*

*Chimonobambusa marmorea* f. *variegata*

*Chimonobambusa marmorea* var. *variegata*

**引证**：*Chimonobambusa marmorea* 'Variegata', Ohwi, Fl. Jap. 2nd Ed.,

1965:135; Ohrnb., The Bamb. World. 181. 1999. ——*Ch. marmorea* (Mitford) Makino f. *variegata* (Makino) Ohwi in Fl. Jap., 75. 1953; Yi et al. in Icon. Bamb. Sin. 272. 2008, et in Clav. Gen. Spec. Bamb. Sin.76.2009; Shi et al. in The Ornamental Bamb. in China. 352. 2012. —— *Ch. marmorea* var. *variegata* (Maki-no) Makino in Bot. Mag. Tokyo 28, 1914: 154. —— *Ch. marrnorea* f. *albovatiegata* Rifat, Nouv. Tahiti, 24 Feb., 1986: 34. ——*Arundinaria marmorea* 'Variegata', A. H. Lawson, Bamb. Gard. Guide, 1968:157. ——*A. marmorea* var. *variegata* Makino in S. Honda, Descr. Prod. For. Jap., 1900: 38; Makino in Bot. Mag.Tokyo 14, 1900: 63.

**特征**：与银明寒竹近似，不同之处在于其秆黄色，节间有几条亮绿色纵条纹，当花青素苷在阳光照射下，其黄色秆还可变为红色；叶片具白色纵条纹。

**用途**：竹姿高雅，极为优美，为著名观赏竹种，尤其适宜用作盆栽、盆景或家居、庭院栽培观赏。

**分布**：日本；中国（江苏南京、四川都江堰）；欧洲；美国。

**（3）方竹** *Chimonobambusa quadrangularis*（Fenzi）Makino

秆高 3~8 m，直径 1~4 cm；节间长 8~22 cm，圆筒形，或下部节间略呈四方形，幼时密被下向黄褐色瘤基小刺毛，毛脱落后留有瘤基而粗糙；箨环初时被黄褐色绒毛及小刺毛；秆环稍平坦或在分枝节上甚隆起；秆中部以下各节内具发达的气生根刺。秆每节上枝条3枚。箨鞘早落，短于节间，背面无毛或有时在中上部贴生极稀疏小刺毛，小横脉紫色，具缘毛；箨耳及箨舌均不发达；箨片锥状，长 3~5 mm。小枝具叶 2~5；叶鞘无毛，鞘口继毛直立；叶舌背面具毛，边缘具细纤毛；叶片长圆状披针形，长 9~29 cm，宽 1~2.7 cm，下面初时被柔毛，次脉 4~7 对。花枝呈总状或圆锥状排列，末级花枝纤细无毛，基部宿存有数片逐渐增大的苞片，具稀疏排列的假小穗 2~4 枚，有时在花枝基部节上即具一假小穗，此时苞片较少；假小穗细长，长 2~3 cm，侧生假小穗仅有先出叶而无苞片；小穗含 2~5 朵小花，有时最下 1 或 2 朵花不孕，而仅具微小的内稃及小花的其他部分；小穗轴节间长 4~6 mm，平滑无毛；颖 1~3 片，披针形，长 4~5 mm；外稃纸质，绿色，披针形或卵状披针形，具 5~7 脉；内稃与外稃近等长；鳞被长卵形；花药长 3.5~4 mm；柱头 2，羽毛状。

1）白纹方竹 *Chimonobambusa quadrangularis* 'Albostriata'

又名：Fuiri-h6chiku，Fuiri-shikaku-dake（日本）

异名：*Chimonobambusa quadrangularis* f. *albostriatus*

*Tetragonocalamus quadrangularis* f. *albostriatus*

引证：*Chimonobambusa quadrangularis* 'Albostriata', Ohrnberger, Bamb. World Gen. Chimonobambusa, 1990:38; J. P. Demoly in Bamb. Assoc. Europ. Bamb. EBS Sect. Fr. No. 8:22, 1991; Ohrnb., The Bamb. World. 184. 1999. ——*Ch. quadrangularis* f. *albostriatus* Stover, Bamb. Book, 1983: 37. —— *Ch. quadrangularis* f. *albostriata* (Muroi & H. Okamura) Wen in J. Bamb. Res. 10 (1): 17, 1991. —— *Tetragonocalamus quadrangularis* f. *albostriatus* Muroi & H. Okamura in Rep. Fuji Bamb. Gard. No. 17, 1972: 10; Muroi in J. Himeji Gakuin Wom. Coll. no. 1: 11, 1974.

特征：与方竹特征相似，不同之处在于其叶具白色条纹。

用途：园艺栽培供观赏。

分布：日本。

2）金纹方竹 *Chimonobambusa quadrangularis* 'Aureostriata'

又名：Kishima-h6chiku，Kishima-shika-ku-dake（日本）

异名：*Chirnonobambusa quadrangularis* f. *aureostriata*

*Tetragonocalamus quadrangularis* f. *albostriatus*

引证：*Chimonobambusa quadrangularis* 'Aureostriata', Ohrnberger, Bamb. World Gen. Chimonobambusa, 1990:38；Ohrnb., The Bamb. World. 184. 1999. ——*Ch. quadrangularis* f. *aureostriata* (Muroi & H. Okamura) Wen in J. Bamb. Res. 10 (1): 17, 1991.——*Tetragonocalamus quadrangularis* f. *albostriatus* Muroi & H. Okamura in Rep. Fuji Bamb. Gard. No. 17, 1972: 10；Muroi in J. Himeji Gakuin Wom. Coll. No. 1: 11, 1974.

特征：与方竹特征相似，不同之处在于其叶具金黄色条纹。

用途：园艺栽培供观赏。

分布：日本。

3）金丝方竹 *Chimonobambusa quadrangularis* 'Cyrano de Bergerac'

异名：*Chirnonobambusa quadrangularis* f. *cyrano-berger-aca*

*Tetragonocalamus quadrangularis* f. *striatus*

**引证**：*Chimonobambusa quadrangularis* 'Cyrano de Bergerac', Rifat in J. Bamb. Res. 6 (2): 25, 1987; Ohrnb., The Bamb. World. 184. 1999. —— *Ch. quadrangularis* f. *cyrano-berger-aca* (Rifat) Wen in J. Bamb. Res. 10 (1): 18, 1991. —— *Tetragonocalamus quadrangularis* f. *striatus* Rifat, ined., ex M. Hirsh, Europ. Bamb. Netw. Newsl. 3, 14, 1986.

**特征**：与方竹特征相似，不同之处在于其秆绿色，但具黄色纵条纹。

**用途**：园艺栽培供观赏。

**分布**：日本；欧洲（从日本引进）。

4）花秆方竹 *Chimonobambusa quadrangularis* 'Joseph de Jussieu'

**又名**：Kimmei-hôchiku, Kimmei-shikaku-dake（日本）

**异名**：*Chimonobambusa quadrangularis* f. *nagaminea*
　　　　*Chimonobambusa quadrangularis* f. *nagamineus*
　　　　*Chimonobambusa quadrangularis* 'Nagamine'
　　　　*Tetragonocalamus quadrangulans* f. *castillonis*
　　　　*Tetragonocalamus quadrangularis* 'Nagamineus'
　　　　*Tetragonocalamus quadrangularis* f. *nagamineus*

**引证**：*Chimonobambusa quadrangularis* 'Joseph de Jussieu', Rifat in J. Bamb. Res. 6 (2) :25, 1987; Ohrnb., The Bamb. World. 185. 1999. —— *Ch. quadrangularis* f. *nagaminea* (Muroi & H. Hamada) Wen in J. Bamb. Res. 10 (1): 18, 1991. —— *Tetragonocalamus quadrangulans* f. *castillonis* Rifat, ined., ex M. Hirsh in Europ. Bamb. Netw. Newsl. 3, 1986:14. —— *T. quadrangulans* 'Nagamineus', Muroi & H. Hamada; cf. H. Okamura in H. Okamura & Y. Tanaka, Hort. Bamb. Sp. Japo, 1986: 31.—— *Tetragonocalamus quadrangularis* f. *nagamineus* Rifat, ined., ex M. Hirsh, Europ. Bamb. Netw. Newsl. 3, 14, 1986; Muroi & H. Hamada ex H. Okamura in H. Okamura & Y. Tanaka, Hort. Bamb. Sp. Jap., 1986: 31.fig. 30, and H. Okamura & M. Konishi in I. c., 1986:89; H. Okamura & al., III. Hort. Bamb. Sp. Jap., 1991:349.——*Ch. quadrangularis* 'Nagamine', Ohrnberger, Bamb. World Gen. Chimonobambusa, 1990: 40, based on *Tetragonocalamus quadrangularis* f. *nagamineus* Muroi & Hamada ex H. Okamura *Chimonobambusa quadrangularis* f.

*nagaminea* G. Bol in Amer. Bamb. Soc. Newsl. 11 (3) : 3, 1990.

**特征**：与方竹特征相似，不同之处在于其秆节间金黄色，具数条浅绿色纵条纹，尤其分枝一侧沟槽为深绿色；叶片偶有白色条纹。

**用途**：珍稀观赏竹类，适于小区、庭院、风景区栽培观赏。

**分布**：日本（本州南部，九州岛，鹿儿岛）；中国。

5）紫秆方竹 *Chimonobambusa quadrangularis* 'Purpureoculmis'

**异名**：*Chimonobambusa quadrangularis* f. *purpureiculma*

**引证**：*Chimonobambusa quadrangularis* (Fenzi) Makino f. *purpureiculma* Wen in Journ. Bamb. Res. 8(1): 24. 1989; Ohrnb., The Bamb. World. 186. 1999; Yi et al. in Icon. Bamb. Sin. 278. 2008, et in Clav. Gen.Spec. Bamb.Sin.79. 2009.

**特征**：与方竹特征相似，不同之处在于其秆为紫色。

**用途**：用于园林绿化，笋可食用。

**分布**：中国（福建顺昌、桂溪）。

6）黄秆方竹 *Chimonobambusa quadrangularis* 'Sotaroana'

**又名**：Gomafu-hôchiku，Gomafu-shikaku-dake（日本）

**异名**：*Chimonobambusa quadrangularis* f. *sotaroana*

*Chimonobambusa quadrangularis* 'Napoleon-Bonaparte'

*Tetragonocalamus quadrangularis* 'Sotaroana'

*Tetragonocalamus quadrangularis* f. *sotaroanus*

*Tetragonocalamus quadrangulans* vat. *sotaroanus*

**引证**：*Chimonobambusa quadrangularis* 'Sotaroana', Ohrnberger, Bamb. World Gen. Chimonobambusa, 1990:38.; Ohrnb., The Bamb. World. 185. 1999. —— *Ch. quadrangularis* f. *sotaroana* (Muroi) Wen in J. Bamb. Res. 10 (1): 17, 1991. ——*Ch. quadrangularis* 'Napoleon-Bonaparte', Rifat in J. Bamb. Res. 6 (2): 25, 1987 ——*Tetragonocalamus quadrangularis* 'Sotaroana', Hatusima, Woody Pl. Jap., 1976. ——*T. quadrangularis* f. *sotaroanus* (Muroi) Muroi in J. Himeji Gakuin Worn. Coll.No. 1, 1974:11.——*T. quadrangulans* vat. *sotaroanus* Muroi in Hyogo Biol. 2, 1948:7.

**特征**：与方竹特征相似，不同之处在于其秆为黄色，偶尔有几条绿色条纹。

**用途**：园艺栽培供观赏。

**分布**：日本（本州南部）；欧洲（1987 由 C. Rifat 从日本引至法国、瑞士）。

7）须尾方竹 *Chimonobambusa quadrangularis* 'Suow'

**又名**： Tatejima-hôchiku，Suow-shikaku- dake（日本）

**异名**：*Chimonobambusa quadrangularis* f. *suhow*

*Chimonobambusa quadrangularis* f. *suou*

*Chimonobambusa quadrangulans* f. *suow*

*Tetragonocalamus quadrangulans* 'Tatejima'

*Tetragonocalamus quadrangulans* f. *tatejima*

**引证**：*Chimonobambusa quadrangularis* 'Suow', Ohrnb., The Bamb. World. 185. 1999.——*Ch. quadrangulans* f. *suow* (Kasahara & H. Okamura) Wen in J. Bamb. Res. 10 (1), 1991: 18, invalid (basionym not validly published) *Chimonobambusa quadrangulans* 'Suow'; Ohrnberger, Bamb. World Gen. Chimonobambusa, 1990: 40. ——*Ch. quadrangularis* f. *suhow* G. Bol in Amer. Bamb. Soc. Newsl. 9 (6): 2, 1988. ——*Ch. quadrangularis* f. *suou* G. Bol in Amer. Bamb. Soc. Newsl. 11 (3) :3, 1990, invalid Chimonobambusa quadrangulans 'Suou'; G. Cooper in Amer. Bamb. Soc. Newsl. 16 (4): 17, 1995. ——*Tetragonocalamus quadrangularis* 'Suow', Kasahara & H. Okamura; cf. H. Okamura in H. Okamura & Y. Tanaka, Hort. Bamb. Sp. Jap., 1986: 31.——*T. quadrangularis* f. *suow* Kasahara & H. Okamura in H. Okamura & Y. Tanaka, Hort. Bamb. Sp. Jap., 1986: 31, fig. 31. ——*T. quadrangulans* 'Tatejima', Kasahara & H. Okamura; cf. H. Okamura in H. Okamura & Y. Tanaka, Hort. Bamb. Sp. Jap., 1986: 31.——*T. quadrangulans* f. *tatejima* Kasahara & H. Okamura ex H. Okamura & al., III. Hort. Bamb. Sp. Jap., 1991: 34,9.

**特征**：与方竹特征相似，不同之处在于其秆为黄色，具一至几条宽窄不同的绿色纵条纹；叶有时会有一些黄白色条纹。

**用途**：园艺栽培供观赏。

**分布**：日本（1980 年发现于日本南部）。美国（1988—1990 年，美国竹子学会从日本引进）。

8）黄槽方竹 *Chimonobambusa quadrangularis* 'Yellow Groove'

**引证**：*Chimonobambusa quadrangularis* 'Yellow Groove', American Bamboo

Society. *Bamboo Species Source List* No. 33：13. Spring 2013.

特征：与方竹特征相似，不同之处在于其秆分枝一侧沟槽为黄色。

用途：园艺栽培供观赏。

分布：不详。

（4）八月竹 *Chimonobambusa szechuanensis*（Rendle）Keng f.

秆高 2.5~4（6）m，直径 1.5~2 cm；节间长 18~22 cm，圆筒形，或下部节间近于四方形，光滑；箨环疏被褐色绒毛；秆环平或稍隆起；秆下部各节内具或多或少的气生根刺。秆每节上枝条 3 枚。箨鞘迟落，短于节间，背面无毛，具紫色纵条纹，具缘毛；箨舌高 0.5~1 mm；箨片锥状三角形，长 3~5 mm。小枝具叶 1~3；叶鞘口繸毛长 3~5 mm；叶舌高 1~1.5 mm；叶片长 18~20 cm，宽 1.2~1.5 cm，次脉 4~6 对。花枝可反复分枝，分枝顶端有叶或无叶，分枝常与假小穗混生于各节上，具叶小枝下部各节有假小穗 1~3 枚；假小穗有 0~4 苞片，上部 1 或 2 片有芽或有次生假小穗；小穗含小花 3 或 4 朵；颖 2 或 3；外稃卵圆披针形，先端渐尖，纵脉 7~9 条；内稃长卵圆形，几与其外稃同长，先端钝圆头或微凹，背部具 2 脊；鳞被 3，靠近外稃的 2 片较近内稃的另一片为大，膜质，上部边缘着生细长白色纤毛；花药黄色；子房卵圆形，花柱甚短，近基部即分裂为 2 枚羽毛状柱头。颖果卵状椭圆形，长 15 mm，粗 6 mm，果皮厚 0.8~1 mm，呈坚果状，但果皮与种皮难于分离，仅与胚乳部分相游离。

1）八月竹 *Chimonobambusa szechuanensis* 'Szechuanensis'

异名：*Chimonobambusa szechuanensis* var. *szechuanensis*

引证：*Chimonobambusa szechuanensis* (Rendle) Keng f. var. *szechuanensis*，Keng et Wang in Flora Reip. Pop. Sin. 9(1): 337. 1996.

特征：与八月竹种特征相似。

用途：生态建设，园林绿化。

分布：中国（四川西部和云南西部）。

2）龙拐竹 *Chimonobambusa szechuanensis* 'Flexuosa'

异名：*Chimonobambusa szechuanensis* f. *flexuosa*

*Chimonobambusa szechuanensis* var. *flexuosa*

引证：*Chimonobambusa szechuanensis* var. *flexuosa* Hsueh et C. Li in

Journ. Yunn. For. Coll. 1982(1): 40. f. 3. 1982; Hsueh et W. P. Zhang in Bamb. Res. 7(3): 10. 1988; Keng et Wang in Flora Reip. Pop. Sin. 9(1): 337. 1996; Yi et al. in Icon. Bamb. Sin. 279. 2008, et in Clav. Gen.Spec. Bamb.Sin.78. 2009.——*Chimonobambusa szechuanensis* f. *flexuosa* (Hsueh et C. Li)Wen et Ohrnb., Gen. Chimonobambusa 44. 1990; Ohrnb., The Bamb. World. 187. 1999.

**特征**：与八月竹特征相似，不同之处在于其秆下部数节有"之"字形强烈曲折并短缩而肿胀的畸形节间。

**用途**：优美的观赏竹种。

**分布**：中国（四川雅安）。

## 2.3 绿竹属 *Dendrocalamopsis*（Chia & H. L. Fung）Keng f.

### （1）线耳绿竹 *Dendrocalamopsis lineariaurita* Yi et L.Yang

地下茎合轴丛生型。秆高 14～15 m，直径 5.5～8 cm，梢头直立；整秆具 35～42 节，节间长 32～38 cm，最长达 42 cm，基部节间长 10～33 cm，圆筒形，分枝一侧无沟槽，平滑，灰绿色，初时被薄白粉质，无毛，中空，竹壁厚 5～12 mm，髓为锯屑状；箨环隆起，褐色，无毛；秆环平，基部数节上具气生根；节内高 4～12 mm，无毛。秆芽扁桃形，具缘毛。秆 4～6 m 开始分枝，每节枝条多枚，主枝长 2.5～3 m，直径达 1 cm，侧枝纤细，直径 2～4 mm，斜展。秆箨早落，厚革质，稍短于节间长度，背面被不均匀贴生棕色刺毛，被薄白粉，稍具纵脉纹，无缘毛；箨耳线形，宽 2～4 mm，灰色；箨舌弧形，灰色，高 2～3 mm，边缘具短而扁的继毛；箨片三角形或长三角形，直立，长 10～25 cm，宽 6～8 cm，背面常被贴生棕色刺毛，边缘具细锯齿。小枝具叶 5～9 枚；叶鞘无毛，上部常被薄白粉，无缘毛；叶耳缺失，鞘口继毛紫褐色或灰褐色，长达 2mm；叶舌弧形，紫褐色，高 1 mm，边缘初时具灰色长达 2 mm 纤毛；叶柄长 2～3 mm，淡绿色，无毛；叶片线形或线状披针形，绿色，纸质，无毛，长 15～25 cm，宽 1.7～3.2 cm，基部截形至楔形，先端长渐尖，次脉 8、9 对，小横脉形成长方形，边缘一侧具小锯齿而粗糙，另一侧近于平滑。花枝未见。笋期 8 月。

本种近似吊丝单 *Dendrocalamopsis vario-striata*（W. T. Lin）Keng f.。不同在于秆节间灰绿色，被薄白粉，梢端直立；箨耳线形或弧形，箨舌较低，高 2.3 mm；叶耳缺失，容易区别。主要用于生态建设、园林绿化。

产中国（四川都江堰）。

1）黄条线耳绿竹 *Dendrocalamopsis lineariaurita* 'Luridilineata'

**异名**：*Dendrocalamopsis lineariaurita* f. *luridilineata*

**引证**：*Dendrocalamopsis lineariaurita* Yi et L.Yang f. *luridilineata* Yi et L.Yang in Journ. Sichuan For. Sci. Techn. 36(3):3. Fig. 6. 2015.

**特征**：与线耳绿竹特征近似，区别在于其秆基部节间具淡黄色纵条纹。

**用途**：园林栽培供观赏。

**分布**：中国（四川都江堰）。

**（2）绿竹 *Dendrocalamopsis oldhami*（Munro）Keng f.**

地下茎合轴型。秆丛生，高 6～12 m，直径 3～9 cm；节间长 20～35 cm，稍"之"字形曲折，幼时被白粉，秆壁厚 4～12 mm。秆分枝高，每节枝条多数，簇生，3 主枝粗壮。箨鞘脱落性，先端近截形，背面无毛或被或疏或密的褐色刺毛，边缘无纤毛或在上部有纤毛；箨耳近等大，椭圆形或近圆形，边缘生纤毛；箨舌高约 1 mm，全缘或波状；箨片直立，三角形，基部截形并收窄，宽度约为箨鞘顶端的 1/2。小枝具叶 6～15；叶鞘初时被小刺毛；叶耳半圆形，继毛棕色；叶舌低矮；叶片长 15～30 cm，宽 3～6 cm，下面被柔毛，次脉 9～14 对，小横脉较清晰，边缘粗糙或有小刺毛。花枝无叶；假小穗下部绿色，上部红紫色，两侧扁，长 2.7～3 cm，宽 7～10 mm，单生或丛生于花枝每节上；苞片 3～5，上方 1 或 2 片腋内无芽；小穗含小花 5～9；小穗轴脱节于颖下；颖片 1，卵形，长 9～10 mm，宽 8 mm，边缘具纤毛，具多脉，有小横脉；外稃卵形，长约 17 mm，宽 13 mm，无毛或有微毛，具约 31 脉，有小横脉，具缘毛；内稃长约 13 mm，两面被毛，顶端尖，背部具 2 脊，脊间具 3～5 脉，脊外两侧各具 2 脉，脉间具小横脉，边缘和脊上具显著纤毛；鳞被 3，卵状披针形，长约 3.5 mm，脉纹明显，边缘具纤毛；雄蕊 6，花丝分离，花药长约 8 mm；子房卵形，长约 2 mm，被粗毛，柱头 3，羽毛状。笋期 5～11 月。花期多在夏、秋季。

1）绿竹 *Dendrocalamopsis oldhami* 'Oldhami'

**又名**：坭竹、石竹、毛绿竹、乌药竹、长枝竹、效脚绿

**异名**：*Dendrocalamopsis oldhami* f. *oldhami*

**引证**：*Dendrocalamopsis oldhami* (Munro) Keng f. f. *oldhami*, Keng et Wang in Flora Reip. Pop. Sin. 9(1): 141. 1996.

**特征**：与绿竹种特征相似。

**用途**：著名笋用竹种，宜鲜食，也可加工制笋干或罐头。笋味美，笋期长，产量高，商品开发价值较大；秆供建筑或劈篾编制竹器，也用于造纸；秆可刮取竹茹用于中药清热除烦。

**分布**：中国（浙江，福建，台湾，广东，广西，海南）。

2）花头黄 *Dendrocalamopsis oldhami* 'Revoluta'

**异名**：*Bambusa oldhami* f. *revoluta*

*Dendrocalamopsis oldhami* f. *revoluta*

*Neosinocalamus revolutus*

*Sinocalamus oldhamii* f. *revolutus*

**引证**：*Dendrocalamopsis oldhami* (Munro) Keng f. f. *revoluta* (W. T. Lin et J. Y. Lin) W. T. Lin in Guihaia 10 (1): 15. 1990; Keng et Wang in Flora Reip. Pop. Sin. 9(1): 142. 1996; Yi et al. in Icon. Bamb. Sin. 181. 2008, et in Clav. Gen.Spec. Bamb.Sin.54. 2009.——*Bambusa oldhami* Munro f. *revoluta* W. T. Lin et J. Y. Lin in Act. Phytotax. Sin. 26 (3): 224. f. 2. 1988; Ohrnb., The Bamb. World. 272. 1999. ——*Neosinocalamus revolutus* (W. T. Lin & J. Y. Lin) Wen in J. Bamb. Res. 10 (1), 1991: 23. ——*Sinocalamus oldhamii* f. *revolutus* (W. T. Lin & J. Y. Lin) W. T. Lin in J. S. China Agr. Univ. 14 (3), 1993:111.

**特征**：与绿竹特征相似，不同之处在于其秆绿色，节间有黄色纵条纹；节内常有一圈灰白色或淡黄白色绢毛环。箨鞘顶端宽，背面仅基部有黄棕色刺毛；箨耳长圆形，常向外翻卷，边缘疏生短纤毛；箨舌高约 1.5 mm，细齿裂；箨片三角形，基部稍外延，与箨耳稍有相连。分枝低，常自基部第三节开始分枝。外稃无小横脉；鳞被近卵形；花柱极短。

**用途**：同绿竹。

**分布**：中国（广东，浙江）。

3）花秆绿竹 *Dendrocalamopsis oldhami* 'Striata'

**异名**：*Dendrocalamopsis oldhami* f. *striata*

**引证**：*Dendrocalamopsis oldhami* (Munro) Keng f. f. *striata* Y. Y. Wang et W. Y. Zhang in Journ. Bamb. Res. 25(1): 26. 2006; Yi et al. in Icon. Bamb. Sin. 181. 2008, et in Clav. Gen. Spec. Bamb. Sin.54. 2009.

**特征**：与绿竹特征相似，不同之处在于其秆节间淡黄色，具绿色纵条纹。箨鞘新鲜时绿色有黄色纵条纹。部分叶片有少量黄色条纹。

**用途**：用于公园、景区栽培观赏。其他同绿竹。

**分布**：中国（浙江瑞安）。

# 3　与竹栽培品种国际登录有关的纸质出版物

2015—2016年正式发表的与竹栽培品种国际登录有关的纸质出版物如下。

1）史军义，靳晓白，2015．国际竹类栽培品种登录权威的建立与发展．栽培植物分类学通讯，(3)：12，13．

2）史军义，王道云，易同培，马丽莎，张学利，姚俊，2015．龙丹竹新品种'花龙丹'．林业科学研究，28(3)：441，442．

3）孙茂盛，史军义，易同培，马丽莎，姚俊，蒲正宇，2015．香竹一新品种'红云'．园艺学报，42(12)：2555，2556．

4）史军义．国际竹类栽培品种登录报告（2013—2014），2015．北京：科学出版社：1-110．

5）姚俊，孙茂盛，史军义，周德群，蒲正宇，杨志杰，2016．乌哺鸡竹一新品种'金殿花竹'．科技信息，(7)：97，98．

6）史军义，周德群，马丽莎，姚俊，蒲正宇，2016．功能竹的定向培育及其可行性．农业科学与技术，17(3)：711-716．

7）史军义，代梅灵，周德群，姚俊，高桂春，2016．方城慈竹一新品种'美菱'．世界竹藤通讯，14(5)：31-33，45．

8）史军义，易同培，周德群，马丽莎，姚俊，2016．国际竹类栽培品种登录的理论与实践．世界竹藤通讯，16(6)：23-28．

9）孙茂盛，史军义，周德群，姚俊，蒲正宇，2016．黄竹新品种'秋实'的选育．北方园艺，(24)：157-159．

# 4 与竹栽培品种国际登录有关的网站建设

## 4.1 网站名称与网址
国际竹类栽培品种登录中心（ICRCB）：
http://www.bamboo2013.org 或 http://www.icrcb.org

## 4.2 网站语言
简体中文、英文。

## 4.3 网站栏目设置及介绍
【首　　页】：集合 ICRCB 简介、信息动态、新登录竹品种常用栏目快捷入口，同时提供国际园艺学会与国际植物栽培品种登录权威及竹类相关的多个网站入口链接。

【中心介绍】：对 ICRCB 的文字介绍及图片展示。

【信息动态】：对外适时发布最新的 ICRCB 工作进展情况及竹类新品种登录最新资讯。

【品种登录】：对外发布已经登录的竹类栽培品种，包含文字介绍及特征图片。

【研究成果】：展示与竹类或国际竹类栽培品种登录相关的著作及论文。

【专家队伍】：对目前国际竹类栽培品种登录委员会的专家简介。

【登　录　园】：对目前国际竹类栽培品种登录园的介绍。

【资料下载】：提供 ICRCB 内部资料（登录申请表、竹类栽培品种国际登录范围等）及国际竹类栽培品种登录相关论文的免费下载服务。

【联系我们】：包含 ICRCB 的地址（提供在线电子地图查询服务）、联系电话、邮编及电子邮箱地址；并提供"留言板"服务。

### 4.4 网站运行情况

网站于2014年10月建成开通,并于2015年7月对网站进行升级改版工作。升级改版工作主要包括:在原有中文语言的基础上,增加了英文服务语言,使网站服务真正进入国际化;网页版面进行了重新设计,比原来更为美观,图片展示更为清晰;开通了手机网站服务功能,让用户随时、随地、随身可通过手机或其他移动设备访问网站,符合目前信息化时代的需求,提供了更方便快捷的用户体验;对信息动态栏目开通了评论入口,让用户即时发表自己阅读资讯之后的感受、观点和意见,形成良好的互动关系;对部分栏目开通了分享功能,针对中国用户使用习惯开通微博、微信、QQ空间等渠道分享,针对国外用户使用习惯开通脸书(facebook)、推特(twitter)渠道分享,使网站内容推广与传播更快更广,聚集具有相似需求的目标用户;开通了在线登录申请,让用户可以更为便捷地进行竹类新品种国际登录申请。

目前网站配备了专业人员对其进行维护,确保24小时正常、全时运行。网站现在除了具有对外发布信息与展示功能外,更兼具了对国内外有相关需求的个人或单位用户提供相关互联服务。

### 4.5 网站服务情况

网站自建成开通以来,为各类个人、单位及社会团体用户提供服务,用户涵盖科研院所用户、高等院校用户、企业用户、政府部门用户、民间组织用户及个人用户,2015—2016年网站累计点击量已超过10000次,资料下载量达1500余次,通过网站首页提供的相关链接进入相应其他网站200余次。

在日益增长的网站访问用户数量的情况下,下一步将计划开通用户注册服务,网站将对用户划分为浏览用户、实名制注册用户和离线用户,分别对不同用户群有针对性地提供相应优质服务;计划针对中文版内容开通微信公众号服务,将原来的网站信息资讯从被动访问方式变成主动向外界推送,从而提高竹类新栽培品种国际登录和信息发布的时效性。

# 5. 国际竹类栽培品种登录园建设

**5.1 国际竹品种（中国·北京）登录园**

国际竹品种（中国·北京）登录园（国际编号：ICRGB-2013-001），英文 International Cultivar Registration Garden for Bamboos（Beijing, China），缩写为 ICRGB（Beijing, China）。该登录园地处中国首都北京，于 2013 年 10 月批准设立，占地面积为 5 hm$^2$，其主要目的是开展北京及其同气候区耐寒观赏竹栽培品种活体植物的收集、保存、国际登录以及优质竹品种的定向培育、展示和推广。

由于人事变动等多种原因，该登录园自设立以来，基本未开展进一步工作。

**5.2 国际竹品种（中国·成都）登录园**

国际竹品种（中国·成都）登录园（国际编号：ICRGB-2014-002），英文 International Cultivar Registration Garden for Bamboos（Chengdu, China），缩写为 ICRGB（Chengdu, China）。该登录园地处中国四川省成都市，于 2014 年 6 月批准设立，占地面积为 12.5 hm$^2$，其主要目的是开展成都及其同气候区各种观赏竹类的收集、保存、国际登录以及优质竹品种的定向培育、展示和推广。

该登录园自设立以来，已收集保存了 5 个竹类栽培品种：

1）斗篷竹 *Neosinocalamus affinis* 'Doupengzhu'
2）佛肚慈竹 *Neosinocalamus affinis* 'Foducizhu'
3）牛腿竹 *Neosinocalamus affinis* 'Niutuizhu'
4）蛇头竹 *Neosinocalamus affinis* 'Shetouzhu'
5）花龙丹 *Dendrocalamus rongchengensis* 'Hualongdan'

**5.3 国际竹品种（中国·都江堰）登录园**

国际竹品种（中国·都江堰）登录园（国际编号：ICRGB-2014-003），

英文 International Cultivar Registration Garden for Bamboos（Dujiangyan，China），缩写为 ICRGB（Dujiangyan，China），该登录园地处中国四川省都江堰市，于 2014 年 8 月批准设立，占地面积为 15 hm$^2$，其主要目的是开展成都及其同气候区各种笋用竹类的收集、保存、国际登录以及优质竹品种的定向培育、展示和推广。

该登录园自设立以来，已收集保存了 2 个竹类栽培品种：

1）都江堰方竹 *Chimonobambusa neopurpurea* 'Dujiangyan Fangzhu'
2）条纹刺黑竹 *Chimonobambusa neopurpurea* 'Lineata'

### 5.4 国际竹品种（中国·昆明）登录园

国际竹品种（中国·昆明）登录园（国际编号：ICRGB-2016-004），英文 International Cultivar Registration Garden for Bamboos（Kunming，China），缩写为 ICRGB（Kunming，China），该登录园地处中国云南省昆明市，于 2016 年 7 月批准设立，占地面积为 26 hm$^2$，其主要目的是开展昆明及其同气候区各种功能竹类的收集、保存、国际登录以及优质竹品种的定向培育、展示和推广。

该登录园自设立以来，已收集保存了 2 个竹类栽培品种：

1）彩云 *Chimonocalamus delicates* 'Caiyun'
2）红云 *Chimonocalamus delicates* 'Hongyun'

### 5.5 国际竹品种（中国·南阳）登录园

国际竹品种（中国·南阳）登录园（国际编号：ICRGB-2016-005），英文 International Cultivar Registration Garden for Bamboos（Nanyang，China），缩写为 ICRGB（Nanyang，China），该登录园地处中国河南省南阳市，于 2016 年 12 月批准设立，占地面积为 10 hm$^2$，其主要目的是开展中原地区及其同气候条件下各种功能竹类的收集、保存、国际登录以及优质竹品种的定向培育、展示和推广。

该登录园自设立以来，已收集保存了 1 个竹类栽培品种：

1）美菱 *Neosinocalamus fangchengensis* 'Meiling'

# 第 2 部分
# 竹类 ICRA 报告
## （2015—2016）
### （英文）

# INTERNATIONAL SOCIETY FOR HORTICULTURAL SCIENCE
# ISHS SPECIAL COMMISSION FOR CULTIVAR REGISTRATION
# ICRA REPORT FOR (2015-2016)

# (Bamboos)

## International Cultivar Registration Center for Bamboos
## December, 2016

Prof. Shi Junyi,
E-mail: esjy@163.com
Web: http://www.bamboo2013.org
Address: Research Institute of Resources Insects, Chinese Academy of Forestry, Bailongsi, Panlong Qu, Kunming, Yunnan 650216, People's Republic of China

# 1 Newly registered bamboo cultivars

## 1.1 'Jindian Huazhu'

*Phyllostachys vivax* 'Jindian Huazhu'

**Applicants:** Sun Maosheng, Yang Zhijie, Xu Honglin, Yi Tongpei, Yao Jun

**Application date:** January 29, 2015

**Preservation locality:** Jindian Scenic Area, Kunming, Yunnan, China

**Authorized date:** March 25, 2015

**Registration No.:** WB-001-2015-009

**Cultivar description:**

Diffuse bamboos. Culms 5-10 m tall, 3-8 cm in diameter, erect, initially green, grey-green when old, with a few light yellow stripes from the base to the middle part of the culm; internodes terete, 25-35 cm long, culm-nodes prominent, a little taller than sheath-nodes, a ring of white powder below the node. Rhizomes green with light yellow stripes. Branching from the upper part of the culm, branches 1-2, usually one branch at the lower part, and two branches at the higher part, branch-nodes prominent. Culm-sheaths papery, deciduous, with dense dark brown patches and spots; auricles and oral setae absent; blades reflexed, strip-lanceolate, strongly crinkled. Leaves 2-3 per ultimate branch, 10-16 cm long, 1.2-2 cm wide. Shooting from the middle to the end of April (Fig. 1-1).

This cultivar was selected from populations of *Ph. vivax* 'Huanwenzhu'.

This cultivar is not only suitable for ornamentation, but also for use of shoots. It is suitable for ornamentation, and shoots are edible.

**Key diagnostic characters compared with the closely related cultivar:**

This cultivar resembles *Phyllostachys vivax* 'Huangwenzhu', but differs in the following characters: culms grey green for the former cultivar, with inconspicuous, narrow and dense, variously wide, light yellow stripes, all over the

Cultivated populations

**Fig. 1-1** *Phyllostachys vivax* **'Jindian Huazhu'**

culm; while culms green for *Phyllostachys vivax* 'Huangwenzhu', with 1-2 golden and broad stripes on the groove (Table 1-1).

Table 1-1  Diagnostic characters of *Phyllostachys vivax* 'Jindian Huazhu' and *Phyllostachys vivax* 'Huangwenzhu'

| Character | *Phyllostachys vivax* 'Jindian Huazhu' | *Phyllostachys vivax* 'Huangwenzhu' |
|---|---|---|
| Culm | Culms grey green | Culms green |
| Stripe colour | Light yellow<br>Inconspicuous | Golden<br>Conspicuous |
| Stripe NO. | A few, narrow and dense,<br>all over the culm | 1-2 on each internode, broad,<br>on the groove only |

## 1.2 'Yanghuang 1'

***Bambusa changningensis*** **'Yanghuang 1'**

**Applicants:** Yi Tongpei, Li Benxiang

**Application date:** June 02, 2015

**Preservation locality:** Bamboo Century Park, Yibin, Sichuan, China

**Authorized date:** July 10, 2015

**Registration NO.:** WB-001-2015-010

**Cultivar description:**

Rhizomes sympodium. Culms 15-19.5 m tall, 8-10 cm in diameter, erect; nodes 41-53, internodes (15) 35-45 (60) cm long, terete, shortly sulcate on the branching side, dark green, glabrous, densely white powdery when young, culm-walls 0.5-1.1 (2.5) cm thick, pith crumby; sheath-nodes prominent, initially green or purple, glossy, glabrous; culm-nodes flat; intranodes 1-1.5 cm long, glabrous, without white powder when young. Culm-buds one, flattened-peach-shaped, glossy, upper margins ciliate. Branching from the $15^{th}$ to $20^{th}$ nodes (6-8 m above the ground), branches many, open, 80-160 (300) cm long, (0.2) 0.5-1.5 cm long in diameter. Shoots dark green, with appressed yellow brown setae at the base; culm-sheaths deciduous, 2/3 as long as the internode, leathery, apex broadly arched, a little asymmetric, usually with appressed brown setae at the base, glossy adaxially, upper margins initially with grey or brown cilia; auricles unequal, oblong, purple, the larger twice as large as the smaller, 1.5 cm long, 0.7 cm wide, oral setae dense, undulate, 7-15 mm long; ligules arched, glabrous, 4-6 mm tall, cilia 2-3 mm long; blades erect, green, triangular, the base 3/4 as wide as the top of the sheath, glabrous abaxially, the adaxial base with sparsely appressed brown setae, margins without cilia. Leaves (4) 6-8 (9) per branchlet; sheaths 7-10 cm long, green, glabrous, ribs obvious, margins without cilia; auricles and oral setae absent; ligules truncate, purple, 1 mm tall, margins with cilia occasionally; petioles 3-5 mm long, light green, glabrous; blades linear-lanceolate, papery, (16) 20-29 (32) cm long, (1.8) 2.4-3 (3.6) cm wide, green adaxially, light green abaxially, apex gradually acute, base broadly wedge-shaped, asymmetric, secondary veins 7-10 pairs, transverse veins ambiguous, margins sawtoothed. Shooting from August to September (Fig. 1-2).

This cultivar was selected through optimization, isolation, colonization, and transplanting from *Bambusa changningensis* Yi et B. X. Li. This cultivar

Fig. 1-2 *Bambusa changningensis* 'Yanghuang 1'

has taller culms, more biomass, and more yield of bamboo wood than *Bambusa changningensis*. The highest weight of a single culm reaches to 32.5 kg, and leaves are 12.5 kg in weight. The average yield of this cultivar is more than one third compared with *Bambusa changningensis*.Culms can be used for papermaking and weaving, and are also important materials for bamboo proximate matters and plywood. This cultivar can be planted for ornamentation. Shoots of this cultivar are sweet and delicate and can be eaten when fresh or dried.

## 1.3 'Zhuomu 1'

*Dendrocalamus mutatus* 'Zhuomu 1'

**Applicants:** Yi Tongpei, Li Benxiang

**Application date:** June 12, 2015

**Preservation locality:** Bamboo Century Park, Yibin, Sichuan, China

**Authorized date:** July 16, 2015

**Registration NO.:** WB-001-2015-011

**Cultivar description:**

Rhizomes sympodium. Culms cespitose, 18-25 m tall, 14-18 cm in diameter, the top erect or a little pendulous; internodes 55-65, 60-70 cm long, basal internodes 15-30 cm long, terete, glabrous, with white powder when young, hollow, culm-walls 0.8-2.5 cm thick, pith crumby; sheath-nodes initially grey, gradually brown, narrow and thin, glabrous; culm-nodes flat, basal nodes with aerial roots; intranodes 1.2-1.6 cm tall, glabrous. Culm-buds flattened, appressed, margins with brown cilia. Branching from 18$^{th}$ to 25$^{th}$ nodes (10-14 m above the ground), dominant branches strong, to 4.5-5 m long, secondary branches several, short and slender. Shoots light purple, blades light yellow-green, reflexed; culm-sheaths decidous, semi-elliptic, thick leathery, shorter than internodes, abaxially with appressed brown setae on the top, ribs inconspicuous, margins without cilia; auricles and oral setae absent; ligules concave, the middle a little higher, purple, 2-5 mm tall, margins fimbriate; blades reflexed, triangular, with appressed brown setae abaxially, margins coarse. Branches with leaves 5-7 (8); sheaths glabrous, upper ribs prominent, green and purple, margins without cilia; auricles and oral setae absent; ligules nearly truncate, purple, glabrous, 1.5 mm tall; petioles 8-10 mm long, glabrous; blades lanceolate, green, papery, glabrous, 24-35 cm long, 4.5-7.5 cm wide, apex gradually acute, the base broadly wedge-shaped, secondary veins 10-13 pairs, transverse veins rectangular, margins sawtoothed. Shooting in August (Fig. 1-3).

This cultivar was selected from populations of *Dendrocalamus mutatus* Yi et B. X. Li. Shoots of this cultivar are delicate, crisp, sweet and rich of nutrition.

Bamboo clump

**Fig. 1-3** *Dendrocalamus mutatus* 'Zhuomu 1'

Bamboo lesves

Spine-aerial roots

Culm-sheath

Young culms

**Fig. 1-3** (continued)

Culms are straight with high biomass. This cultivar is suitable for the use of shoots and culms. This cultivar has taller culms, more biomass, and more yield of bamboo wood than *Dendrocalamus mutatus*. The culms can reach to 22 m tall and more than 17 cm in diameter.

## 1.4 'Qiushi'

*Dendrocalamus membranaceus* 'Qiushi'

**Applicants:** Sun Maosheng, Tang Chengsong, Duan Shengbiao

**Application date:** October 08, 2015

**Preservation locality:** Bamboo Garden of Southwest Forestry University, Kunming, Yunnan, China

**Authorized date:** November 18, 2015

**Registration NO.:** WB-001-2015-012

**Cultivar description:**

Cespitose bamboos. Culms 10-18 m tall, 6-10 (13) cm in diameter, erect; internodes terete, 20-30 cm long, white powdery initially; culm-nodes flat, sheath-nodes prominent; internodes from the basal first to the 18$^{th}$ with several golden stripes; the basal three nodes with aerial roots; branching from the basal nodes; dominant branches 3. Culm-sheaths leathery, deciduous, longer than internodes at the base, shorter than internodes at the upper, with brown and green stripes abaxially when fresh, glabrous, the top truncate; ligules 5-8 mm tall, lobed; auricles absent or tiny, with dense brown tomenta; blades reflexed, linear-lanceolate, 5-33 cm long, 2.5-4 cm wide, with brown tomenta abaxially, especially dense at the base. Leaves 4-8 per branchlet; ligules inconspicuous, 1 mm tall; auricles falcate, oral setae purple; blades lanceolate, 11.5-22 cm long, 10-15 mm wide, the base wedge-shaped, white tomentose on both surfaces. Shooting from late June to early September (Fig.1-4).

This cultivar was selected from the wild population of *Dendrocalamus membranaceus* Munro.

This cultivar has great value of ornamentation and can be planted in gardens and for bonsai, bamboo corridors, fences, and so on. It is suitable for use of culms and shoots and a valuable bamboo cultivar with high quality and multi-function. This cultivar with great exploitation potentiality can be planted widely in suitable areas.

**Key diagnostic characters compared with the closely related cultivar:**

This cultivar resembles *Dendrocalamus membranaceus* 'Striatus', but differs

Bamboo clump  Bamboo shoot  Bamboo culms

**Fig. 1-4** *Dendrocalamus membranaceus* 'Qiushi'

in several characters. Culms of *Dendrocalamus membranaceus* 'Qiushi' have several conspicuous golden stripes from the basal first to the 18$^{th}$ internodes even upper; conspicuous brown or green stripes also occur on fresh shoots or culm-sheaths; culm-sheaths are glabrous. Culms of *Dendrocalamus membranaceus* 'Striatus' have 1-2 narrow golden stripes on the basal 1-2 internodes; fresh shoots or culm-sheaths possess several inconspicuous brown and green stripes; culm-sheaths are covered with brown tomenta abaxially (Table 1-2).

**Table 1-2  Diagnostic characters of *Dendrocalamus membranaceus* 'Qiushi' and *Dendrocalamus membranaceus* 'Striatus'**

| Character | *Dendrocalamus membranaceus* 'Qiushi' | *Dendrocalamus membranaceus* 'Striatus' |
|---|---|---|
| culm | Several different-width golden stripes present from the basal to the middle internodes | Golden stripes present only on the basal 1-2 internodes |
| culm-sheath | Glabrous abaxially<br>Brown and green stripes conspicuous | Brown tomenta present on abaxial surface<br>No tripes or inconspicuous |

## 1.5 'Qicai'

***Bambusa lapidea* 'Qicai'**

**Applicants:** Sun Maosheng, Tang Chengsong, Duan Shengbiao

**Application date:** November 18, 2015

**Preservation locality:** Bamboo Garden of Southwest Forestry University, Kunming, Yunnan, China

**Authorized date:** December 25, 2015

**Registration No.:** WB-001-2015-013

**Cultivar description:**

Cespitose bamboos. Culms 8-18 m tall, 6-12 cm in diameter, straight at the base, zigzag at the top; culm-walls thick; internodes terete, 22-35 cm long, green initially, yellow later; culms with golden stripes from the base to the middle part; basal nodes with aerial roots; branching from the basal nodes; dominant branches 3. Culm-sheaths thick leathery, deciduous at the upper part of the culm, tardily deciduous at the base of the culm, shorter than internodes, with light yellow stripes when fresh, glabrous; blades erect, broadly ovate-triangular, the base shrinked inside, glabrous abaxially, brown setose adaxially; ligules 2-5 mm tall, lobed, margins ciliate; auricles conspicuous, unequal, wrinkled ovate, with dense brown tomenta. Leaves 4-12 per branchlet, linear-lanceolate, 11-26 cm long, 1.5-3 cm wide, the base round, glabrous. Shooting from late June to early August (Fig. 1-5).

This cultivar was selected from populations of *Bambusa lapidea* McClure.

This cultivar has great value of ornamentation and can be planted in gardens and for bonsai, bamboo corridors, and so on. It is suitable for use of culms and shoots and a valuable bamboo cultivar with high quality and multi-function.

Cultivated populations　　　　　　　　Bamboo shoot

Bamboo culm　　　　　　　　Bamboo leaves

**Fig. 1-5　*Bambusa lapidea* 'Qicai'**

## 1.6 'Zhuhai Yingtouhuang'

*Bambusa rigida* 'Zhuhai Yingtouhuang'

**Applicants:** Chen Qibing, Jiang Mingyan, Lü Bingyang, Jiang Tao, Li Nian

**Application date:** March 02, 2016

**Preservation locality:** Shuguang Cultivar Bamboo Garden, Changning, Yibin, Sichuan, China

**Authorized date:** April 12, 2016

**Registration No.:** WB-001-2016-014

**Cultivar description:**

Cespitose bamboos. Culms 10-14 m tall, 4-9 cm in diameter; nodes flat, young culms with white wax, glabrous, 25-35 nodes, internodes 30-50 (60) cm long; dominant branches prominant, leaves 5-12 per terminate branchlets, rectangular, 12-25 cm long, 2-3 cm wide, adaxially dark green, glabrous, pink green abaxially; old culms with 1-3 dormant buds. Shooting from July to September (Fig. 1-6).

*Bambusa rigida* 'Zhuhai Yingtouhuang' is selected from populations of *Bambusa rigida* Keng et Keng f. Culms of this cultivar are straight and 6.5-7.5 cm (9.2 cm) in diameter. The yield of bamboo culms is high with single culms to 18 kg (the heaviest to 28 kg). The weight of an individual culm and the yield per acre are 30% to 50% more than the other cultivars of *B. Rigida*. Culms can be used for poles, scaffolds, farming tools, and paper making. Bamboo shoots are edible. The rate of survival from cuttage is more than 80%.

This cultivar can be planted in various habitats, such as barren mountains, village edges, roadsides, and riversides.

**Key diagnostic characters compared with *Bambusa rigida*:**

(1) This cultivar grows in broader distribution areas. It can be planted at the altitude below 800 m in subtropical and southern subtropical areas.

(2) It has taller and thicker culms and thicker culm-walls. The biomass of this cultivar promotes significantly, and the yield of culms is much higher.

(3) More new bamboo shoots occur each year than *Bambusa rigida*, and the yield of culms per unit area is more than *Bambusa rigida*.

Cultivated populations

Culm-sheath　　　　　　　　　　　　Bamboo leaves

Fig. 1-6　*Bambusa rigida* 'Zhuhai Yingtouhuang'

(4) This cultivar can resist to massive water and chilliness to some extent. The capability of cold resistence is similar to *Lingnania intermedia* (Hsueh et Yi) Yi.

## 1.7 'Huayetangzhu'

*Sinobambusa tootsik* 'Huayetangzhu'

**Applicants:** Chen Songhe, Wang Zhenzhong

**Application date:** March 10, 2016

**Preservation locality:** Xiamen Botanical Garden, Xiamen, Fujian, China

**Authorized date:** April 12, 2016

**Registration No.:** WB-001-2016-015

**Cultivar description:**

Diffuse bamboos. Culms 5-12 m tall, 2-6 cm in diameter; internodes 30-40 (80) cm, with white powder initially, especially dense below nodes, ribs present on old culms, flat and sulcate on the branching side; sheath-nodes corky, with purple-brown setae initially; culm-nodes prominent, as tall as sheath-nodes. Branches usually 3, ocassionally 5-7, branch nodes prominent. Culm-sheaths deciduous, nearly retangular, the top obtuse, green when fresh, with yellow-white stripes, stripes especially wide at the edge; auricles oblong to ovate, falcate at the top of the culm, tomentose or coarse, oral setae waved, to 2 cm long; ligules 4 mm tall, arched, maigins ciliate or glabrous; blades green, lanceolate or long lanceolate, reflexed, margins sparsely serrate. Leaves 3-6 (9) per branchlet; auricles inconspicuous, oral setae radiate, to 15 mm long; ligules 1-1.5 mm; blades 6-22 cm long, 1-3.5 cm wide, with yellow-white stripes (Fig. 1-7).

This cultivar was selected from the individuals with striped leaves of *Sinobambusa tootsik* (Sieb.) Makino. Culm-sheaths are green with yellow-white stripes, especially wide at the edge, and leaves with yellow-white stripes as well. This cultivar is suitable for planting in yards and gardens.

Fig. 1-7 *Sinobambusa tootsik* 'Huayetangzhu'

## 1.8 'Meiling'

*Neosinocalamus fangchengensis* 'Meiling'

**Applicants:** Dai Meiling, Ma Lisha, Gao Guichun, Yao Jun, Zhai Dehua

**Application date:** June 18, 2016

**Preservation locality:** International Cultivar Registration Gardens for Bamboos (Nanyang · China)

**Authorized date:** August 09, 2016

**Registration No.:** WB-001-2016-016

**Cultivar description:**

Cespitose bamboos. Culms 12-14 m tall, 3-4 cm in diameter, the top pendulous; internodes 41-43, 40-44 cm long, the basal internodes 20 cm long, terete, initially with white powder, internodes at the middle and base of the culm with light yellow stripes, culm-walls thin, 2-4 (5) mm thick; sheath-nodes prominent, purple brown, glabrous; culm-nodes flat; intranodes 2-3 mm tall, glabrous, with white powder. Buds almond-shaped, glabrous. Branching from upper nodes, branches many, spreading, dominant branches to 1.7 m long, 4-5 mm in diameter, lateral branches slender, 1-2 mm in diameter. Shoots light green, with sparse brown setae and yellow stripes. Culm-sheaths deciduous, leathery, shorter than internodes, with sparse brown setae and yellow stripes; auricles absent; ligules truncate, initially purple, 2-2.5 mm tall, oral setae dense, flat, initially purple brown, 5-12 mm long; blades linear or linear-triangular, 8-18 cm long, reflexed. Leaves 8-13 per branchlet; sheaths purple green to green, glabrous; auricles and oral setae absent; ligules arched, purple brown, glabrous, 1.5 mm tall; petioles light green, glabrous, 1.5-2 mm long; blades linear-lanceolate, 16-25 cm long, 2.6-4 cm wide. Shooting from July to August (Fig. 1-8).

This cultivar was selected from the individuals of *Neosinocalamus fangchengensis* Yi et J. Y. Shi which have light yellow stripes on the culms. Bamboo shoots and fresh culm-sheahs are also with brown and green stripes. This cultivar has great value of ornamentation. It grows well at the temperature -5℃, and it can be planted at the similar climate in the northern and central regions of China, especially regions near Han River, for ornamentation and bonsai.

Bamboo clump

**Fig. 1-8** *Neosinocalamus fangchengensis* 'Meiling'

Fig. 1-8 (continued)

## 1.9 'Aijiaoma'

*Dendrocalamus latiflorus* 'Aijiaoma'

**Applicants:** Chen Songhe, Huang Kefu

**Application date:** August 01, 2016

**Preservation locality:** Xiamen Botanical Garden, Xiamen, Fujian, China

**Authorized date:** October 19, 2016

**Registration No.:** WB-001-2016-017

**Cultivar description:**

Cespitose bamboos. Culms less than 12 m, less than 15 cm in diameter, internodes less than 15 cm long; branching from the basal fifth or sixth nodes, the basal three nodes usually with aerial roots; leaves large, papery. Shooting from early May to early November, the shooting summit from the early July to early August; the weight of a single shoot light, approximately 1-3 kg, but with high yield per unit area due to long shooting period (Fig. 1-9).

This cultivar was selected from populations of *Dendrocalamus latiflorus*

Bamboo clump　　　　　　　　Bamboo culm

**Fig. 1-9** *Dendrocalamus latiflorus* 'Aijiaoma'

Bamboo shoot  Culm-sheaths

Bamboo leaves  Young bamboos

**Fig. 1-9** (continued)

Munro which have more delicious bamboo shoots. This cultivar has been widely planted in southern Fujian. However, this cultivar does not resist to coldness, and it is suitable to be cultivated in southern subtropical areas.

## 1.10 'Luaijiao'

*Dendrocalamopsis oldhami* 'Luaijiao'

**Applicants:** Chen Songhe, Huang Kefu

**Application date:** August 01, 2016

**Preservation locality:** Xiamen Botanical Garden, Xiamen, Fujian, China

**Authorized date:** October 19, 2016

**Registration No.:** WB-001-2016-018

**Cultivar description:**

Cespitose bamboos. Culms usually 4-6 m tall, 4.5-6.0 cm in diameter, internodes less than 30 cm, branching from the nodes 0.8-1.0 m above the ground. Shooting from middle or late May to late October, the shooting summit from the middle to late July. The base of the bamboo shoot is inclined as a horse hoof, therefore, it is usually called as horse-hoof bamboo shoot. The bamboo shoots are small and tapering, but it has a high yield per unit area due to long shooting period (Fig. 1-10).

Bamboo clump　　　　　　　　Bamboo culm

Fig. 1-10　*Dendrocalamopsis oldhami* 'Luaijiao'

Bamboo shoots

**Fig. 1-10** (continued)

This cultivar selected from populations of *Dendrocalamopsis oldhami* (Munro) Keng f. which have more delicious bamboo shoots. The bamboo shoots of this cultivar are crisp and delicious. This cultivar has been planted widely in southern Fujian. The culms can be used for paper making. This cultivar does not resist to the coldness, and it is suitable to be planted in southern subtropical area.

## 1.11 'Lineata'

*Chimonobambusa neopurpurea* 'Lineata'

**Applicants:** Ma Lisha, Yin Xianxiao, Li Zhiwei, Liu Yutao, Yao Jun

**Application date:** October 06, 2016

**Preservation locality:** International Cultivar Registration Gardens for Bamboos (Dujiangyan · China)

**Authorized date:** October 29, 2016

**Registration No.:** WB-001-2016-019

**Cultivar description:**

Culms 2-5 m tall, 1-3 cm in diameter; internodes 10-16 (20) cm long, green, slightly quadrate; the basal internodes light purple, with light green stripes; sheath-nodes with dense brown setae initially; culm-nodes a little prominent. Branches 3. Culm-sheaths persistent, shorter than internodes; auricles absent, oral setae absent or several; ligules arched; blades erect, subulate, 1-2 mm long. Leaves 2-4 per branchlet; oral setae absent, or several; blades linear-lanceolate, 5-18 cm long, 0.5-2 cm wide, glabrous or sometimes with grey yellow pubescence at the base abaxially, secondary veins 4-6 pairs, transverse veins conspicuous (Fig. 1-11).

This cultivar was selected from cultivated populations of *Chimonobambusa neopurpurea* Yi which have colorful stripes on the culms. This cultivar has great value of ornamentation, and it can be planted in gardens or as bonsai. Bamboo shoots are edible and delicious.

**Key diagnostic characters compared with related species:**

*Chimonobambusa neopurpurea* 'Lineata' is closely related to *Ch. neopurpurea* 'Dujiangyan Fangzhu', and the key diagnostic characters are as follows: the basal internodes light purple with light green stripes, culm-sheaths shorter than the internode, bamboo shoots darker for the former cultivar; the basal internodes green without stripes, culm-sheaths longer than the internode, bamboo shoots lighter for the latter one (Table 1-3).

Bamboo clump  Culm-sheaths

Bamboo leaves  Bamboo shoot

**Fig. 1-11** *Chimonobambusa neopurpurea* **'Lineata'**

Table 1-3　Diagnostic characters of *Chimonobambusa neopurpurea* 'Lineata' and *Ch. neopurpurea* 'Dujiangyan Fangzhu'

| Bamboo taxa | 'Lineata' | 'Dujiangyan Fangzhu' |
|---|---|---|
| Culms | Basal young culms light purple with light green stripes | Basal culms green without stripes |
| Culm-sheaths | Shorter than internodes | Longer than internodes |
| Bamboo shoot | Darker | Lighter |

## 1.12 'Chuanmuzhu'

*Dendrocalamus mutatus* 'Chuanmuzhu'

**Applicants:** Chen Qibing, Jiang Mingyan, Lv Bingyang, Li Nian, Cen Huameng, Zhang Cheng

**Application date:** November 01, 2016

**Preservation locality:** Bamboo Garden, Xinya Street, High-tech Zone, Chengdu, Sichuan, China

**Authorized date:** November 15, 2016

**Registration No.:** WB-001-2016-020

**Cultivar description:**

Cespitose bamboos. Culms 15-20 m tall, 8-16 cm in diameter; nodes prominent, basal nodes with aerial roots, slightly white powdery, a ring of dense white powder present below young sheath-nodes. The longest internodes present above the ground 5-6 m, usually 45-55 cm long, the longest to 65 cm. Branching from basal nodes, branches many; leaves 6-10 per branchlet; leaves narrowly lanceolate, 12-30 cm long, 1.5-3 cm wide, adaxially coarse, abaxially pubescent, margins serrated, secondary veins 3-6 pairs. Culm-sheaths deciduous, thick-papery, the top arched, auricles tiny. The culm wall of this cultivar is thick, and the thickest is at the base of the culm with the average thickness 4 cm and gradually thinner with the mean thickness at the middle culm 1.9-2.5 cm. The average weight of the individual culm is 60 kg, even to 86 kg. Bamboo shoots are edible, and the shooting period is during July to September (Fig. 1-12).

This cultivar was selected from individuals of *Dendrocalamus mutatus* Yi et B. X. Li which have thick culm-walls. The culms of this cultivar can be used for various use, and the bamboo shoots are edible. It is also suitable for ornamentation in yards and gardens.

**Key diagnostic characters compared with related species:**

*Dendrocalamus mutatus* 'Chuanmuzhu' is closely related to *Dendrocalamus mutates* 'Zhuomu 1'. The key diagnostic characters are as follows: the former

Bamboo clump

**Fig. 1-12** *Dendrocalamus mutatus* 'Chuanmuzhu'

| Bamboo culms | Culm-sheaths | Bamboo shoots |

**Fig. 1-12** (continued)

cultivar with thicker culm-walls at the base ( to 4 cm thick) and gradually thinner, the average thickness 1.9-2.5 cm at the middle culm, mean weight 60 kg (to 86 kg); the latter with thinner culm-walls.

# 2 Summarization of the published bamboo cultivars

The registrar of the International Cultivar Registration Authority for Bamboos surveyed many literatures during 2015-2016 in order to find those published bamboo cultivars in the past years. After careful collation, 73 cultivars of 18 bamboo species in *Bambusa* Retz. corr. Schreber, *Chimonobambusa* Makino, and *Dendrocalamopsis* (Chia & H. L. Fung) Keng f. were recorded here. There are 55 cultivars of 12 species in *Bambusa*, including two cultivars of *Bambusa blumeana*, three cultivars of *Bambusa dissimulator*, two cultivars of *Bambusa dolichoclada*, three cultivars of *Bambusa eutuldoides*, 22 cultivars of *Bambusa multiplex*, two cultivars of *Bambusa pervariabilis*, one cultivars of *Bambusa rigida*, 10 cultivars of *Bambusa textilis*, one cultivar of *Bambusa tulda*, three cultivars of *Bambusa tuldoides*, two cultivars of *Bambusa ventricosa*, four cultivars of *Bambusa vulgaris*; fourteen cultivars of four species in *Chimonobambusa*, including one cultivar of *Chimonobambusa angustifolia*, three cultivars of *Chimonobambusa marmorea*, eight cultivars of *Chimonobambusa quadrangularis*, two cultivars of *Chimonobambusa szechuanensis*; four cultivars of two species, including one cultivars of *Dendrocalamopsis lineariaurita*, three cultivars of *Dendrocalamopsis oldhami*. All the bamboo cultivars were listed alphabetically.

## 2.1 *Bambusa* Retz. corr. Schreber

### (1) *Bambusa blumeana* J. A. et J. H. Schult.F.

Culms 15-24 m tall, 8-15 cm in diameter, the top pendulous, the base a little zigzag; internodes 25-35 cm long, the upper with sparse, brown, and procumbent setae initially, culm-walls 2-3 cm thick; the lower intranodes with short aerial roots, a ring of grey-white or brown tomenta above and below sheath-nodes. Branching from the basal first node, branchlets of the lower branches usually specialized into tough thorns forming dense clusters, the middle and upper nodes with three or more than three branches, usually with dominant branches. Culm-sheaths tardily deciduous, with densely dark-brown setae abaxially, ribs prominent, the top broadly arched or concaved, shoulders protruded; auricles equal or a little unequal, reflexed, oral setae undulate and long; ligules 4-5 mm tall, fimbriate; blades reflexed, ovate or narrowly ovate, with densely dark-brown setae adaxially, the base narrowed and then extened laterally to be auricles, basal margins with cilia, the base 2/5 as broad as the top of culm-sheaths. Branchlets with leaves 5-9; the upper of leaf-sheaths with setae, margins ciliate; auricles tiny or absent, oral setae absent or 2-3; ligules truncate, fimbriate; blades 10-20 cm long, 1.2 -2.5 cm wide, glabrous, the base with dense pubescence abaxially. Two or several pseudospikelets clustered on nodes, linear, light-purple, 2.5-4 cm long, 3-4 mm wide, 4-12 florets for each pseudospikelet, 2-5 florets bisexual; glumes 2, 2 mm long, glabrous; lemma ovate-oblong, 6-9 mm long, 2.5-4 mm wide, glabrous, 9-11 veins, apex acute, margins without cilia; palea 7 mm long, 1.8 mm wide, 2-keeled with dense cilia, 3 veins between and outside keels, respectively, apex dichotomous; filaments free, 6-7 mm long, anthers yellow, broadly linear, 3-4 mm long; ovary bottle-shaped, 1.2-2 mm long, style short, stigmas 3, plumose. Shooting from June to September, flowering usually in spring (occasionally in middle November).

1) *Bambusa* 'Blumeana'

**Local names:** Lezhu、Yuzhu (China); Bambu duri (Indonesia); Pring gesing, Shi-chiku(Japan); Buloh duri, Buloh sikai (Malaysia); Phai-si-suk, Mai-si-suk (Thailand); Kauayan tinik, Bata-kan, Kawayan-siitan (the Philippines); Rüssèi roliëk (Cambodia); Phaix ba:nz (Laos); Tre gai (Vietnam).

**Citations:** *B.blumeana* cv. Blumeana, Keng et Wang in Flora Reip. Pop. Sin. 9(1): 53. 1996.

**Characteristics:** The same with *Bambusa blumeana*.

**Use:** This bamboo can be planted as ecological shelter forests; culms can be used as scaffolds.

**Distribution:** China (Fujian, Taiwan, Guangxi, Yunnan); Indonesia (Java Island); eastern Malaysia;the Philippines; Thailand; Vietnam; Cambodia; Laos.

2) *Bambusa blumeana* 'Wei-fang Lin'

**Local names:** Huifang Lezhu, Lin's Lezhu (China); Rinshi-chiku (Japan)

**Synonyms:** *Bambusa blumeana* f. *wei-fang lin*

*Bambusa stenostachya* cv. Wei-fan Lin

**Citations:** *Bambusa blumeana* 'Wei-fang Lin', Ohrnb., The Bamb. World. 257. 1999.——*B. blumeana* J. A. et J. H. Schult. cv. Wei-fang Lin (W. C. Lin) Chia et al. in Guihaia 8(2) : 123. 1988; Keng et Wang in Flora Reip. Pop. Sin. 9(1): 54. 1996.——*B. stenostachya* Hack. cv. Wei-fan Lin W. C. Lin in Bull. Taiwan For. Res. Inst. No. 98: 12. f. 7-8, 1964; Fl. Taiwan 5: 761. 1978.——*B. blumeana* J. A. et J. H. Schult. f. *wei-fang lin*(W. C. Lin) Yi in Journ. Sichuan For. Sci. Techn. 28(3): 17. 2007; Yi et al. Icon. Bamb. Sin. 97. 2008, et in Clav. Gen. Spec.Bamb.Sin. 29. 2009; Shi et al. in The Ornamental Bamb. in China. 277. 2012.

**Characteristics:** Culms and branches golden, gradually becoming orange with dark green stripes; culm-sheaths light green with several yellow stripes.

**Use:** This bamboo can be cultivated in yards and gardens for ornamentation.

**Distribution:** China (northern Taiwan).

**(2) *Bambusa dissimulator* McClure**

Culms 10-18 m tall, 4-7 cm in diameter, the base a little zigzag; internodes 25-35 cm long, with thin white powder when young, glabrous, sometimes the basal nodes with yellow-white stripes; the basal first and second intranodes with aerial roots occasionally. Branch solitary on the first or second node, the upper nodes with 3 to many branches clustered, dominant branches strong and long, branchlets of the basal branches usually specialized into thorns. Culm-sheaths deciduous, nearly glabrous abaxially, or with inconspicuous setae, ribs prominent, the top asymmetrically arched; auricles unequal, wrinkled, oral setae undulate, the larger one oblong to oblanceolate, 4-5 mm wide, the smaller one ovate to oblong, 3-4 mm wide; ligules 5-7 mm tall, lobed, with white fimbria; blades erect, ovate-triangular to ovate-lanceolate, with brown setae among veins adaxially, the base narrowed roundly or heart-shaped, 3/5 to 1/2 as wide as the top of culm-sheaths, margins with undulate setae at the base. Leaf-sheaths glabrous, or coarse, or with dense setae; auricles absent or inconspicuous, oral setae several; ligules short, truncate, entire; blades 7-18 cm long, 10-18 mm wide, with sparse short pubescence abaxially, especially dense along the middle rib. Pseudospikelets single or several clustered on nodes, lanceolate, flat, ca. 3 cm long; prophyll with 2 keels; 2 bracts with buds, ovate, apex blunt; rachilla 2-3 mm long, the surface opposite to palea flat and glabrous, the surface on the other side coarse, the top swollen with cilia; pseudospikelets with 4 or 5 fertile florets, the top with 2 to several sterile florets; glume 1 or sometimes absent, like lemma; lemma lanceolate, to 12 mm long, glabrous, apex blunt or acute with subulate tip, upper margins sometimes with short cilia; palea with 2 keels, keels strongly folded near the apex and with short cilia or coarse, the apex of palea usually with a cluster of pubescence, 5 veins between keels; lodicules 3, nearly equal, ovate or obovate, apex blunt or a little concaved, margins with long cilia; filaments free, apex of anthers blunt, a little concaved; ovary obovate or ovate, pediculate, the top thickened and with setae; style single, short, tomentose, stigmas 3. Shooting from July to August. Flowering from March to April.

1) *Bambusa* 'Dissimulator'

**Local names:** Nilezhu, Lezhu, Nizhu, Zhunafu (China)

**Synonyms:** *Bambusa dissimulator* var. *dissimulator*

**Characteristics:** The same with *Bambusa dissimulator*.

**Citations:** *Bambusa* 'Dissimulator', Shi et al. in World Bamb. Ratt. 14(6): 27. 2016.——*B. dissimulator* McClure var. *dissimulator*, Keng et Wang in Flora Reip. Pop. Sin. 9(1): 60. 1996.

**Use:** This bamboo can be cultivated as fences, and culms can be used as scaffolds.

**Distribution:** China (Guangdong, Guangxi).

2) *Bambusa dissimulator* 'Albinodia'

**Local names:** Baijielezhu (China)

**Synonyms:** *Bambusa dissimulator* var. *albinodia*

**Citations:** *Bambusa dissimulator* 'Albinodia', Shi et al. in World Bamb. Ratt. 14(6): 27. 2016.——*B. dissimulator* var. *albinodia* McClure in Lingnan.

Sci. Journ. 19(3): 415. 1940; Flora of Guangzhou 774. 1956; Chen, Illustr. manual of Chinese trees and shrubs (supplement). 11. 1957; But et al., Bamboos in Hongkong 31. 1985; Keng et Wang in Flora Reip. Pop. Sin. 9(1): 61. 1996; Ohrnb., The Bamb. World. 259. 1999; Flora of China. 22: 14. 2006; Yi et al. Icon. Bamb. Sin. 99. 2008, et in Clav. Gen. Spec.Bamb.Sin. 31. 2009.

**Characteristics:** A ring of pale white tomenta present above and below sheath-nodes.

**Use:** This bamboo can be used as scaffolds.

**Distribution:** China (Guangdong, Hongkong).

3) *Bambusa dissimulator* 'Hispida'

**Local names:** Maolezhu (China)

**Synonyms:** *Bambusa dissimulator* var. *hispida*

**Citations:** *Bambusa dissimulator* 'Hispida', Shi et al. in World Bamb. Ratt. 14(6): 27. 2016.——*B. dissimulator* var. *hispida* McClure in Lingnan Sci. Journ. 19 (3): 415. 1940; Flora of Guangzhou. 774. 1956; Chen, Illustr. manual of Chinese trees and shrubs (supplement) 11. 1957; Keng et Wang in Flora Reip. Pop. Sin. 9(1): 61. 1996; Ohrnb., The Bamb. World. 259. 1999; Flora of China. 22: 14. 2006; Yi et al. in Icon. Bamb. Sin. 99. 2008, et in Clav. Gen. Spec.Bamb. Sin. 31. 2009.

**Characteristics:** Nodes and internodes of the culm with coarse setae; culm-sheaths with setae abaxially.

**Use:** This bamboo can be used as scaffolds.

**Distribution:** China (Guangdong).

**(3) *Bambusa dolichoclada* Hayata**

Culms 10-15 m, 4.5-8 cm in diameter; internodes 30-45 cm long, with thick white powder initially, glabrous, a ring of grey-white tomenta above basal sheath-nodes. Branching from the basal first node, 3 to many branches clustered, dominant branches strong and long. Culm-sheaths deciduous, with thin white powder abaxially, the top and upper parts with dense brown setae, the top asymmetrically arched, or sometimes nearly truncate, upper margins with cilia; auricles unequal, a little wrinkled, apex blunt round, oral setae undulate, the larger ovate-oblong or narrowly ovate, 2-2.5 cm long, 8-10 mm wide, the smaller one ovate or elliptical, 1/3 as large as the larger one; ligules 3-4 mm tall, lobed, macrame hairs to 5 mm long; blades erect, deciduous, asymmetrically ovate-triangular, with brown setae abaxially, with thin white powder initially, with dense light-brown setae adaxially, the base narrowed roundly and linked with auricles, the linkage 3-5 mm, 2/3 as wide as the top of culm-sheaths, lower margins with cilia. Leaf-sheaths with setae abaxially; auricles narrowly ovate, oral setae to 8 mm long; ligules with setae abaxially; blades 10-26 cm long, 1-2.3 cm wide, with pubescence abaxially. Pseudospikelets 3-9 clustered on nodes, linear, 3-4 cm long, 6-8 mm wide, 4-12 florets for each pseudospikelet, subtended by several bracts with buds; glumes 2, ovate or broadly ovate, 2-4.5 mm long, veins 14, apex acute; lemma ovate, 9 mm long, veins 18-20 with transverse veins, apex acute; palea 8.5 mm long, 2 keels with dense cilia; anthers yellow, 4.5 mm long, apex concave; ovary obovate, 2 mm long, the top with pubescence, style very short, stigmas 3, plumose.

1) *Bambusa* 'Dolichoclada'

**Local names:** Tongzaizhu (China); Choshi-chiku (Japan); Long-Branch Bamboo, Long-Shoot Bamboo (English)

**Citations:** *Bambusa* 'Dolichoclada', Shi et al. in World Bamb. Ratt. 14(6): 27. 2016. ——*B.dolichoclada* cv. Dolichoclada Hayata, Keng et Wang in Flora Reip. Pop. Sin. 9(1): 94. 1996.

**Characteristics:** The same with *Bambusa dolichoclada*.

**Use:** This bamboo can be planted as ecological shelter forests;culms can be used for buildings, furniture making, and farming tools.

**Distribution:** China (Fujian, Taiwan); Japan (Okinawa, Kyushu).

2) *Bambusa dolichoclada* 'Stripe'

**Synonyms:** *Bambusa dolichoclada* f. *stripe*

*Bambusa ventricosa* f. *stripe*

**Citations:** *Bambusa dolichoclada* 'Stripe', Ohrnb., The Bamb. World. 260. 1999; American Bamboo Society. *Bamboo Species Source List* No. 33: 7. Spring 2013. ——*B. dolichoclada* Hayata cv. Stripe W. C. Lin in Bull. Taiwan For. Res. Inst. No.98: 15. f. 9, 10. 1964; Fl. Taiwan 5: 751. 1978; Keng et Wang in Flora Reip. Pop.

Sin. 9(1): 96. 1996.——*B. ventricosa* Hayata f. *stripe* (W. C. Lin) Yi;Yi et al. Icon. Bamb. Sin. 118. 2008, et in Clav. Gen.Spec. Bamb.Sin. 37. 2009; Shi et al. in The Ornamental Bamb. in China. 288. 2012.

**Characteristics:** Culms shorter and slender than *Bambusa dolichoclada*, golden, with green stripes; leaves green with light yellow and white stripes.

**Use:** This bamboo can be cultivated as bonsai, or in the yards and gardens for ornamentation.

**Distribution:** Southern China and Taiwan; Japan (Kyushu).

### (4) *Bambusa eutuldoides* McClure

Culms 6-12 m tall, 4-6 cm in diameter; internodes 30-40 cm long, with thin white powder when young or not, glabrous or the upper parts with setae, a ring of grey white tomenta above and below sheath-nodes, culm-walls 5 mm thick. Branching from the basal second or third node, branches many, 3 dominant branches strong and long. Culm-sheaths deciduous, leathery, glabrous or with sparse procumbent setae abaxially, occasionally with yellow white strips near the outer margins, the outer shoulder extended downward, asymmetrically arched; auricles unequal, crisp, wrinkled, oral setae present, the larger one extended downward 2/5 to 1/2 as long as the culm-sheaths, oblanceolate or narrowly oblong, 5-6.5 cm long, 1.5 cm wide, the smaller one nearly round or oblong, 1 cm in diameter, or inconspicuous; ligules 3-5 mm tall, lobed, fimbriate; blades erect, asymmetrically triangular or narrowly triangular, the base narrowed and extended to link with auricles, the linkage 1 cm, with sparse setae abaxially, the adaxial base with brown setae, the base 3/5 as wide as the top of culm-sheaths. Leaf-auricles present or not, oral setae several; ligules 0.5 mm tall, truncate, lobed; blades 12-25 cm long, 1.4-2.5 cm wide, with dense pubescence abaxially. Pseudospikelets several to many clustered on nodes, linear, 2.5-5.5 cm long; 5-6 florets for each floret, subtended by several bracts with buds; rachilla flat, 3-4 mm long, the top swollen with cilia; glume 1, oblong, 9-10 mm long, glabrous, with purple spots abaxially, veins 11, apex acute with a cusp; lemma 12-13 mm long, veins 13-15; palea lanceolate, 11 mm long, 2 keels, the top with short cilia, 4 veins between keels, 2 veins outside each keel; lodicules 3, unequal, margins with long cilia, the former two narrow, 2 mm long, the back one large, broadly obovate or nearly round, 2 mm long; anthers 5 mm long, apex split; ovary nearly spherical, 1 mm in diameter, the top with short setae, style very short with setae, stigmas 3, plumose. Caryopsis nearly obovate, 5 mm long, the top with setae and residue of style.

1) *Bambusa* 'Eutuldoides'

**Local names:** Dai Ngan Bamboo (English)

**Synonyms:** *Bambusa eutuldoides* var. *eutuldoides*

**Citations:** *Bambusa* 'Eutuldoides', Shi et al. in World Bamb. Ratt. 14(6): 27. 2016.——*B. eutuldoides* var. *eutuldoides*, Keng et Wang in Flora Reip. Pop. Sin. 9(1): 84. 1996.

**Characteristics:** The same with *Bambusa eutuldoides*.

**Use:** Culms can be used for buildings, farming tools, and weaving.

**Distribution:** China (Guangdong, Guangxi, Hongkong).

2) *Bambusa eutuldoides* 'Basistriata'

**Local names:** Bannizhu (China)

**Synonyms:** *Bambusa eutuldoides* var. *basistriata*

**Citations:** *Bambusa eutuldoides* 'Basistriata', Shi et al. in World Bamb. Ratt. 14(6): 27. 2016.——*B. eutuldoides* McClure var. *basistriata* McClure in Lingnan Univ. Sci. Bull. No. 9: 9. 1940; Bamboos in Guangxi and cultivation, 36, 1987; Keng et Wang in Flora Reip. Pop. Sin. 9(1): 85. 1996; Ohrnb., The Bamb. World. 261. 1999; Flora of China 22: 22. 2006; Yi et al. Icon. Bamb. Sin. 121. 2008, et in Clav. Gen.Spec. Bamb. Sin. 36. 2009; Shi et al. in The Ornamental Bamb. in China. 285. 2012.

**Characteristics:** Basal internodes with yellow and white stripes; culm-sheaths with yellow and white stripes abaxially, the larger auricles strongly wrinkled.

**Use:** This bamboo can be planted for ornamentation.

**Distribution:** China (Guangxi, Guangdong).

3) *Bambusa eutuldoides* 'Viridi-vittata'

**Synonyms:** *Bambusa eutuldoides* var. *viridi-vittata*

*Bambusa eutuldoides* var. *viridivittata*

**Citations:** *Bambusa eutuldoides* 'Viridi-vittata', American Bamboo Society. *Bamboo Species Source List* No. 33: 7. Spring 2013.——*B. eutuldoides* McClure var. *viridi-vittata* McClure in Lingnan Univ. Sci. Bull. No. 9: 9. 1940; Bamboos in Guangxi and cultivation, 36, 1987; Keng et Wang in Flora Reip. Pop. Sin. 9(1): 85. 1996. ——*B. eutuldoides* var. *viridivittata*, Ohrnb., The Bamb. World. 261. 1999; Flora of China 22: 22. 2006; Yi et al. Icon. Bamb. Sin. 121. 2008, et in Clav. Gen. Spec. Bamb.Sin. 37. 2009; Shi et al. in The Ornamental Bamb. in China. 285. 2012.

**Characteristics:** Internodes yellow with green stripes; culm-sheaths green with yellow stripes, the larger auricles shorter and strongly wrinkled.

**Use:** This bamboo can be planted for ornamentation.

**Distribution:** China (Guangdong, Jiangxi, Sichuan).

**(5) *Bambusa multiplex* (Lour.) Raeuschel ex J. A. & J. H. Schult.**

Culms 4-6 (7) m, 2-4 cm in diameter; internodes green, 30-50 cm long, with thin white powder when young, the upper with brown setae, culm-walls thin; nodes a little prominent, glabrous. Branching from basal second to third nodes, many branches clustered, dominant branches strong and long. Culm-sheaths tardily deciduous, with thin white powder abaxially initially, glabrous, the top asymmetrically arched; auricles tiny or inconspicuous, oral setae several; ligules 1-1.5 mm tall, lobed; blades erect, deciduous, long-triangular, with dark-brown setae abaxially, coarse adaxially, the base as wide as the top of culm-sheaths. Branchlets with 5-12 leaves; auricles kidney-shaped, oral setae undulate; ligules 0.5 mm tall, lobed; blades 5-16 cm long, 0.7-1.6 cm wide, the abaxial epidermis pale green with dense grey white pubescence. Pseudospikelets single or several clustered on nodes, subtended by sheath-like bracts, linear to linear lanceolate, 3-6 cm long; prophyll 3.5 mm long, 2 keels with short cilia; 1 or 2 bracts with buds, ovate to narrowly ovate, 4-7.5 mm long, glabrous, 9-13 veins, apex blunt or acute; (3) 5-13 florets for each pseudospikelet, the middle florets bisexual; rachilla flat, 4-4.5 mm long, glabrous; glume absent; lemma asymmetrically, oblong-lanceolate, 18 mm long, glabrous, veins 19-21, apex acute; palea linear, 14-16 mm long, 2 keels with cilia, 6 veins between keels, 4 veins outside one keel and

3 veins outside the other keel, apex with 2 cusps, the top truncate with cilia; the former two lodicules semi-ovate, 2.5-3 mm long, the back one long lanceolate, 3-5 mm long, margins glabrous; filaments 8-10 mm long, anthers purple, 6 mm long, apex with a cluster of pubescence; ovary ovoid, 1 mm long, the top thickened with setae, petiole 1 mm long, stigmas 3 or varied, 5 mm long, plumose.

1) *Bambusa* 'Multiplex'

**Local names:** Huochuizhu, Huoguanzhu, Huoguangzhu, Huokaizhu (China); Hourai-chiku, Houou-chiku (Japan); Buloh Cina, Buloh pagar (Malaysia); Mai-liang, Mai-phai-lieng (Thailand); Bambu cina (Indonesia); Kawayan tsina, Kawayan sa sonsong (the Philippines); Cay hop (Vietnam); Hedge Bamboo (English)

**Synonyms:** *Bambusa multiplex* var. *multiplex*

**Citations:** *Bambusa* 'Multiplex', Shi et al. in World Bamb. Ratt. 14(6): 27. 2016.——*B. multiplex* var. *multiplex*, Keng et Wang in Flora Reip. Pop. Sin. 9(1): 109. 1996.

**Characteristics:** The same with *Bambusa multiplex*.

**Use:** This bamboo can be planted for ornamentation and fences.

**Distribution:** China (southern to southwestern China, Taiwan); Vietnam; Thailand; Indonesia; the Philippines.

2) *Bambusa multiplex* 'Albostriata'

**Local names:** Fuiri-houou (Japan); Silverstripe Fernleaf Hedge Bamboo (English)

**Synonyms:** *Bambusa glaucescens* f. *albosttiata*

*Barnbusa multiplex* f. *albostriata*

*Bambusa multiplex* var. *multiplex*

**Citations:** *Bambusa multiplex* 'Albostriata', Ohrnb., The Bamb. World. 268. 1999.——*B. multiplex* f. *albostriata* Muroi & Sugimoto ex Muroi in J. Himeji Gakuin Wom. Coll. No. 1, 1974:1.——*B. glaucescens* f. *albosttiata* Muroi & Sugimoto. 9,1971——*B. multiplex* var. *multiplex*, Keng et Wang in Flora Reip. Pop. Sin. 9(1): 109. 1996.

**Characteristics:** Basal internodes green with several yellow and white stripes.

**Use:** This bamboo can be planted for ornamentation.

**Distribution:** Japan (the middle part).

3) *Bambusa multiplex* 'Alphonse-Karr'

**Local names:** Suhou-chiku (Japan); Alphonse Karr Hedge Bamboo (English)

**Synonyms:** *Bambusa alphonso-Karri*

*Barnbusa alphonso-karrii*

*Bambusa glaucescens* 'Alphonse Karr'

*Barnbusa glaucescens* 'Alphonso-Karrir'

*Barnbusa glaucescens* f. *alphonso-karrii*

*Bambusa glaucescens* f. *alphonso-karri*

*Bambusa multiplex* 'Alphonso-Karr'

*Bambusa multiplex* 'Alphonso-Karri'

*Bambusa multiplex* 'Alphonso-Karrii'

*Bambusa multiplex* f. *alphonso-Karri*

*Barnbusamultiplex* f. *alphonso-karrii*

*Bambusa multiplex* var. *normalis*

*Barnbusa multiplex* var. *normalis* f. *alphonso-karrii*

*Bambusa nana* var. *normalis* f. *alphonso-karrii*

*Bambusa nana* f. *alphonso-karri*

*Bambusa nana* var. *alphonso karri*

*Barnbusa nana* var. *alphonso-karrii*

*Bambusa nana* var. *normalis*

*Leleba multiplex* f. *alphonso-karri*

*Lelebamultiplex* f. *alphonso-karrii*

**Citations:** *Bambusa multiplex* 'Alphonse-Karr', American Bamboo Society. *Bamboo Species Source List* No. 33: 8. Spring 2013.——*B. multiplex* (Lour.) Raeusch. cv. Alphonse-Karr R. A. Young in USDA Agr. Handb. No.193: 40. 1961; Keng et Wang in Flora Reip. Pop. Sin. 9(1): 112. 1996. ——*B. multiplex* 'Alphonso-Karrii', Ohrnb., The Bamb. World. 267. 1999. —— *B. multiplex* f. *alphonso-karri* (Satow) Nakai in Rika kyoiku 15: 67. 1932; Yi et al. in Icon. Bamb. Sin. 128. 2008, et in Clav. Gen.Spec. Bamb. Sin. 39. 2009; Shi et al. in The Ornamental Bamb. in China. 290. 2012.——*B. alphonso-Karri* Mitf. ex Satow in Trans Asiat. Soc. Jap. 27: 91. pl. 3. 1899; Takeuchi Yoshio, Research on bamboos (in Chinese). 110. 1957.——*B. alphonso-karrii* Mitford, Bamb. Gard.,: 55, 216. 1896. ——*B. nana* var. *alphonsokarri* (Satow) Marliac ex E. G. Camus, Bambus. 121. 1913.——*B. nana* var. *norrnalis* f. *alphonso-karrii* Makino in S. Honda, Descr. Prod. For. Jap., 1900: 37.——*B. nana* f. *alphonso-karrii* (Mitford ex Satow) Makino ex Kawamura, 1907: 2. ——*B. nana* Roxb. var. *normalis* Makino ex Shirosawa f. *alphonso-karri* (Mitf. ex Satow) Makino ex Shirosawa, Icon. Bamb. Jap. 56. pl. 9. 1912. ——*B. multiplex* var. *normalis* Sasaki f. *alphonso-karri* Sasaki, Cat. Gov. Herb. (Form.) 68. 1930.——*Leleba multiplex* (Lour.) Nakai f. *alphonso-karri* (Satow) Nakai in Journ. Jap. Bot. 9: 14. 1933.——*Bambusa multiplex* f. *alphonso-Karri* (Mitf.) Sasaki ex Keng f. in Techn. Bull. Nat'l. For. Res. Bur. China No. 8: 17. 1948; Flora Illustr. Plant. Prima. Sinica. Gramineae. 57, pl. 39. 1959; Bamboos in Guangxi and cultivation, 40, pl. 22. 1987. ——*B. multiplex* 'Alphonse Karr'; R. A. Young in Nation. Hort. Mag. 25, 1946: 260, 264. ——*B. glaucescens* (Lam.) Munro ex Merr. f. *alphonso-karri* (Satow) Hatusima, Fl. Ryukyus .854. 1971.——*B. glaucescens* 'Alphonso-Karrir'; Hatusima, Woody Pl. Jap., 1976: 316. ——*B. glaucescens* (Willd.) Sieb. ex Munro cv. Alphonse Karr (Young) Chia et But in Photologia 52(4): 258. 1982. —— *B. glaucescens* 'Alphonse Karr'; Crouzet, 1981: 51. ——*B. glaucescens* f. *alphonso-karri* (Mitf.) Wen in Journ. Bamb. Res. 4(2): 16. 1985.

**Characteristics:** Culms and internodes yellow with green stripes; culm-sheaths green with yellow white stripes when fresh. Leaves with several yellow and white stripes occasionally. Freeze resistant, to -10℃.

**Use:** This bamboo can be planted for ornamentation.

**Distribution:** China (Sichuan, Guangdong, Taiwan); Japan; Europe; USA (Florida); cultivated in tropical countries (Southern Asia, Southeastern Asia, Eastern Asia).

4) *Bambusa multiplex* 'Fernleaf'

**Synonyms:** *Bambusa floribunda*

*Bambusa glaucescens* cv. Fernleaf

*Bambusa multiplex* f. *fernleaf*

*Bambusa multiplex* var. *elegans*

*Bambusa multipex* var. *fernleaf*

*Bambusa multiplex* var. *nana*

*Bambusa nana* var. *gracillima*

*Ischurochloa floribunda*

*Leleba elegans*

*Leleba floribunda*

**Citations:** *Bambusa multiplex* 'Fernleaf', American Bamboo Society. *Bamboo Species Source List* No. 33: 8. Spring 2013.——*B. multiplex* (Lour.) Raeusch. cv. Fernleaf R. A. Young in USDA Agr. Handb. No. 193: 40. 1961; Fl. Taiwan. 5: 755. 1978; Keng et Wang in Flora Reip. Pop. Sin. 9(1): 113. 1996. ——*Ischurochloa floribunda* Buse ex Miq., Fl. Jungh. 390. 1851. —— *Bambusa floribunda* (Buse) Zoll. et Maur. ex Steud., Syn. Pl. Glum. 1: 330. 1854. ——*B. nana* Roxb. var. *gracillima* Makino ex E. G. Camus, Bambus. 121. 1913, non Kurz 1866. ——*Leleba floribunda* (Buse) Nakai in Journ. Jap. Bot. 9: 10. pl. 1. 1933. ——*L. elegans* Koidz. in Act. Phytotax. Geobot. 3: 27. 1934.——*Bambusa multipex* var. *fernleaf* R. A. Yung in Nat'l. Hort. Mag. 25: 261. 1946. ——*B. multiplex* var. *nana* (Roxb.) Keng f. in Techn. Bull. Nat'l. For. Res. Bur. China No.8: 17. 1948, non *B. nana* Roxb. 1832; Flora Illustr. Plant. Prima. Sinica. Gramineae. 57, pl. 38. 1959; Bamboos in Guangxi and cultivation, 41, pl. 23, 1987. —— *B. multiplex* var. *elegans* (Koidz.) Muroi ex Sugimoto, New Keys Jap. Trees. 457. 1961; S. Suzuki. Ind. Jap. Bambusas. 104, 105 (pl.18), 340. 1978. ——*B. glaucescens* (Willd.) Sieb. ex Munro cv. Fernleaf (R. A. Young) Chia et But in Phytologia. 52(1): 258. 1982; But et al., Bamboos in Hongkong 38. 1985; Chia et al., Chinese bamboos 22. 1988.——*B. multiplex* (Lour.) Raeuschel ex J. A. et J. H. Schult. f. *fernleaf* (R. A. Young) Yi in Yi et al. Icon. Bamb. Sin. 129. 2008, et in Clav. Gen.Spec. Bamb. Sin. 40. 2009; Shi et al. in The Ornamental Bamb. in China. 291. 2012.

**Characteristics:** Culms 3-6 m tall; branchlets pendulous, leaves 9-13, pinnate; leaves 3.3-6.5cm long, 4-7 mm wide.

**Use:** This bamboo can be planted for ornamentation or bonsai.

**Distribution:** China (eastern, southern, and southwestern China, Taiwan, Hongkong).

5) *Bambusa multiplex* 'Floribunda'

**Local names:** Houou-chiku (Japan); Fernleaf Hedge Bamboo (English)

**Synonyms:** *Barnbusa elegans*

*Bambusa glaucescens* f. *elegans*

*Bambusa multiplex* var. *elegans*

*Bambusa nana* var. *disticha*

*Leleba elegans*

**Citations:** *Bambusa multiplex* 'Floribunda', Ohrnb., The Bamb. World. 269. 1999. ——*B. multiplex* var. *elegans* (Koidzumi) Muroi in Sugimoto, New Keys Jap. Tr., 1961:457. ——*Bambusa multiplex* 'Wang Tsai', S. Dransfield & E. A. Widjaja, Pl. Resources S. E. Asia, 7, 1995: 66. ——*B. elegans* Koidzumi ex Murata in Kitamura & Murata, Col. Il1. Woody Pl. Jap., 2, 1979: 369.——*B. glaucescens* f. *elegans* (Koidzumi) Muroi & Sugimoto ex Muroi & H. Okamura, Take sasa, 1977:147, 66.——*B. nana* var. *disticha* hort. ex R. A. Young in Nation. Hort. Mag. 25, 1946: 261.

**Characteristics:** Resemble *Bambusa multiplex* 'Fernleaf'. Culms short; leaves inconspicuously pinnate, clustered, slender, toward the top of the branchlets.

**Use:** This bamboo can be planted for ornamentation.

**Distribution:** Japan (Honshu); Europe; USA; China.

6) **Bambusa multiplex** 'Golden Goddess'

**Synonyms:** *Bambusa glaucescens* 'Golden Goddess'

*Bambusa glaucescens* var. *lutea*

*Bambusa multiplex* var. *lutea*

**Citations:** *Bambusa multiplex* 'Golden Goddess', S. Dransfield & E .A. Widjaja, Pl. Resources S.E. Asia, 7, 1995:66; Ohrnb., The Bamb. World. 267. 1999; American Bamboo Society. *Bamboo Species Source List* No. 33: 8. Spring 2013.

——*B. glaucescens* 'Golden Goddess', Haubrich, 2, 1981. ——*B. multiplex* var. *lutea* Wen, 1982:31. ——*Bambusa glaucescens* var. *lutea* (Wen) Wen, 1985:16.

**Characteristics:** Culms short, 3-3.1 m tall, 1-1.3 cm in diameter; internodes golden; leaves large.

**Use:** This bamboo can be planted for ornamentation.

**Distribution:** Europe; USA (Florida); China.

7) *Bambusa multiplex* 'Incana'

**Synonyms:** *Bambusa multiplex* var. *incana*

*Bambusa strigosa*

**Citations:** *Bambusa multiplex* (Lour.) Raeuschel ex J. A. et J. H. Schult. var. *incana* B. M. Yang, Nat. Sci. Journ. Hunan Norm. Univ. 1983 (1): 77. f. 1. 1983; Keng et Wang in Flora Reip. Pop. Sin. 9(1): 110. 1996; Yi et al. Icon. Bamb. Sin. 127. 2008.——*Bambusa strigosa* Wen in Journ. Bamb. Res. 1(1): 31. pl. 8. 1982.

**Characteristics:** Culm-sheaths with setae abaxially.

**Use:** This bamboo can be planted for ornamentation.

**Distribution:** China (Jiangxi, Hunan).

8) *Bambusa multiplex* 'Kimmei-Suhou'

**Local names:** KJmmei-suhou (Japan)

**Synonyms:** *Bambusa glaucescens* f. *kimrnei-suhou*

*Barnbusa multiplex* f. *kimmei-suhou*

**Citations:** *Bambusa multiplex* 'Kimmei-Suhou', Ohrnb., The Bamb. World. 268. 1999. ——*B. glaucescens* f. *kimrnei-suhou* Muroi & Ka- sahara, 1972:7. ——*Barnbusa multiplex* f. *kimmei-suhou* Muroi & Kasahara ex Muroi in J. Himeji Gakuin Wom. Coll. No. 1, 1974.

**Characteristics:** Culms yellow, occasionally with several green narrow stripes.

**Use:** This bamboo can be planted for ornamentation.

**Distribution:** Japan.

9) *Bambusa multiplex* 'Midori'

**Local names:** Midori-hou-shiyou (Japan)

**Synonyms:** *Bambusa glaucescens* f. *midori*

*Barnbusa glaucescens* f. *alphonso-karrii* 'Midori'

*Bambusa glaucescens* 'Midori'

*Bambusa multiplex* f. *midori*

**Citations:** *Bambusa multiplex* 'Midori' , Ohrnb., The Bamb. World. 267. 1999.——*B. glaucescens* f. *midori* Muroi & Sugimoto, 10, 1971.——*B. multiplex* f. *midori* Muroi & Sugimoto ex Muroi in J. Himeji Gakuin Wom. Coll. no. 1, 1974: 2, as syn.——*B. glaucescens* 'Midori', Stover. 34.1983.——*Barnbusa glaucescens* f. *alphonso-karrii* 'Midori'; Muroi & Sugimoto ex H. Okamura & M. Konishi in H. Okamura & Y. Tanaka, Hort. Bamb. Sp. Jap., 95.1986.

**Characteristics:** Culms 4-4.6 m tall, 3-3.8 cm in diameter; internodes yellow with green stripes.

**Use:** This bamboo can be planted for ornamentation.

**Distribution:** Japan.

10) *Bambusa multiplex* 'Midori Green'

**Citations:** *Bambusa multiplex* 'Midori Green', American Bamboo Society. *Bamboo Species Source List* No. 33: 8. Spring 2013.

**Characteristics:** Resemble *Bambusa multiplex* 'Midori'. Culms and branches bright green with dark green stripes.

**Use:** This bamboo can be planted for ornamentation.

**Distribution:** Japan.

11) *Bambusa multiplex* 'Pubivagina'

**Synonyms:** *Bambusa multiplex* var. *pubivagina*

**Citations:** *Bambusa multiplex* (Lour.) Raeuschel ex J. A. et J. H. Schult. var. *pubivagina* W. T. Lin et Z. J. Feng in Journ. Bamb. Res. 12(2): 33.1993; Yi et al. in Icon. Bamb. Sin. 127. 2008.

**Characteristics:** Internodes green with white stripes;culm-sheaths with densely white or brown tomenta.

**Use:** This bamboo can be planted for ornamentation.

**Distribution:** China (Pingyuan, Wuzhishi of Guangdong).

12) *Bambusa multiplex* 'Riviereorum'

**Local names:** Riviere Hedge Bamboo (English)

**Synonyms:** *Bambusa glaucescens* var. *riviereorum*

*Bambusa multiplex* var. *nana*

*Bambusa multiplex* var. *riviereorum*

*Bambusa sciptoria*

**Citations:** *Bambusa multiplex* 'Riviereorum', Ohrnb., The Bamb. World. 269. 1999; American Bamboo Society. *Bamboo Species Source List* No. 33: 9. Spring 2013. ——*B. multiplex* (Lour.) Raeuschel ex J. A. et J. H. Schult. var. *riviereorum* R. Maire, Fl. Afr. Nord. 1: 355. 1952;Flora of China 22: 31. 2006; Yi et al. Icon. Bamb. Sin. 130. 2008, et in Clav. Gen.Spec. Bamb.Sin. 39. 2009; Shi et al. in The Ornamental Bamb. in China. 290. 2012.——*B. glaucescens* (Willd.) Sieb. ex Munro var. *riviereorum* (R. Maire) Chia et H. L. Fung in Phytologia 52(4): 257. 1982; But et al., Bamboos in Hongkong. 39. 1985.——*B. multiplex* (Lour.) Raeusch. var. *nana* auct. non (Roxb.) Keng f.: Flora of Guangzhou. 771. 1956.——*B. sciptoria* auct. non Dentist.: A. et C. Riv. in Bull. Soc. Acclim. III. 5: 685. 1878.

**Characteristics:** Resemble *Bambusa multiplex* 'Fernleaf'. Culms short, 1-3 m tall, 3-5 mm in diameter, internodes solid; leaves 13-23 for each branchlet, pendulous, 1.6-3.2 cm long, 2.5-6.5 mm wide. Freeze resistant, to -8℃.

**Use:** This bamboo can be planted for ornamentation or as fences.

**Distribution:** China (central and southwestern parts); Indonesia; Thailand; Europe; Africa.

13) *Bambusa multiplex* 'Shimadai'

**Synonyms:** *Bambusa glaucescens* var. *shimadai*

*Bambusa multiplex* var. *shimadai*

*Bambusa shimadai*

*Leleba shimadai*

**Citations:** *Bambusa multiplex* (Lour.) Raeuschel ex J. A. et J. H. Schult. var. *shimadai* (Hayata) Sasaki in Trans. Nat. Hist. Soc, Form. 21: 118. 1931; Keng et Wang in Flora Reip. Pop. Sin. 9(1): 110. 1996; Yi et al. Icon. Bamb. Sin.

127. 2008. ——*Bambusa shimadai* Hayata, Icon. Pl. Form. 6: 151. f. 59. 1916. Takeuchi Yoshio, Researchon bamboos (in Chinese) 110. 1957; W. C. Lin in Bull. Taiwan For. Res. Inst. No. 98; 24. ff. 15, 16. 1964. —— *Leleba shimadai* (Hayata) Nakai in Journ. Jap. Bot. 9(1): 17. 1933; W. C. Lin in 1. c. No.69: 65. ff. 30, 31. 1961. —— *Bambusa glaucescens* (Willd.) Sieb. ex Munro var. *shimadai* (Hayata) Chia et But in Phytologia 52(4): 258. 1982.

**Characteristics:** The top of culm-sheaths nearly symmetrically arched.

**Use:** This bamboo can be planted for ornamentation or as fences and shelter forest.

**Distribution:** China (Guangdong, Taiwan).

14) *Bambusa multiplex* 'Shirosuji'

**Local names:** Shirosuji-kama, Shirosuji-bakama (Japan)

**Synonyms:** *Bambusa glaucescens* 'Shirosuji'

*Bambusa glaucescens* f. *shirosuji*

*Bambusa multiplex* f. *shirosuji*

**Citations:** *Bambusa multiplex* 'Shirosuji', Ohrnb., The Bamb. World. 268. 1999. ——*B. multiplex* f. *shirosuji* Muroi & H. Okamura ex Muroi in J. Himeji Gakuin Wom. Coll. No. 1, 1974: 2. ——*B. glaucescens* 'Shirosuji', Stover, 1983:34. ——*B. glaucescens* f. *shirosuji* Muroi & H. Okamura, 1972:7.

**Characteristics:** Internodes with slender white stripes.

**Use:** This bamboo can be planted for ornamentation.

**Distribution:** Japan.

15) *Bambusa multiplex* 'Shyokomachi'

**Local names:** Shiyou-komachi (Japan)

**Synonyms:** *Bambusa glaucescens* f. *shyokomachi*

**Citations:** *Bambusa multiplex* 'Shyokomachi', Ohrnb., The Bamb. World. 267. 1999. ——*B. glaucescens* f. *shyokomachi* Muroi & Maruyama ex Murci & H. Okamura, Take sasa, 1977: 149, 69.

**Characteristics:** Leaves blue and green.

**Use:** This bamboo can be planted for ornamentation.

**Distribution:** Japan.

16) *Bambusa multiplex* 'Silverstripe'

**Synonyms:** *Bambusa dolichomerithalla* 'Silverstripe'

*Bambusa floribunda* f. *albo-variegata*

*Bambusa glaucescens* cv. Silverstripe

*Bambusa multiplex* 'Albovariegata'

*Bambusa multiplex* f. *silverstripe*

*Bambusa multipex* var. *elegans* f. *albo-varegata*

*Bambusa multiplex* var. *silverstripe*

*Bambusa nana*. var. *albo-variegata*

*Leleba floribunda* f. *albo-variegata*

**Citations:** *Bambusa multiplex* 'Silverstripe', American Bamboo Society. *Bamboo Species Source List* No. 33: 9. Spring 2013.——*B. multiplex* (Lour.) Raeusch. cv. Silverstripe R. A. Young in USDA Agr. Handb. No. 193: 41. 1961; Keng et Wang in Flora Reip. Pop. Sin. 9(1): 112. 1996. ——*B. dolichomerithalla* 'Silverstripe', Lin in Bull. Taiwan For. Res. Inst. No.271: 1976; ——*B. multiplex* 'Albovariegata', Ohrnb., The Bamb. World. 266. 1999. ——*B. multiplex* (Lour.) Raeuschel ex J. A. et J. H. Schult. f. *silverstripe* (R. A. Young) Yiin Yi et al. Icon. Bamb. Sin. 130. 2008, et in Clav. Gen.Spec. Bamb.Sin. 39. 2009; Shi et al. in The Ornamental Bamb. in China. 290. 2012. ——*B. nana* f. *albo-variegata* Makino in Journ. Jap. Bot. 1: 28. 1917. ——*B. nana* Roxb. var. *albo-variegata* E. G. Camus, Bambusa. 121. 1913. ——*B. floribunda* (Buse) Zoll. et Maur. ex Sieb. f. *albo-variegata* Nakai in Riko Kyoiku. 15: 66. 1932. ——*Leleba floribunda* (Buse)Nakai f. *albo-variegata* Nakai in Journ. Jap. Bot. 9: 12.1933.——*B. multiplex* var. *silverstripe* R. A. in Nat'l. Hort. Mag. 25: 260, 264. 1946.——*B. multipex* var. *elegans* (Koidz.) Muroi f. *albo-varegata* (Makino) Muroi ex Sugimoto, New Keys Jap. Trees. 457. 1961. ——*B. glaucescens* (Willd.) Sieb. ex Munro cv. Silverstripe (R. A. Young) Chia et But in Phytologia 52(4): 259. 1982; But et al., Bamboos in Hongkong 40, 1985.

**Characteristics:** Basal internodes, fresh culm-sheaths, and several leaves with white stripes.

**Use:** This bamboo can be planted for ornamentation.

**Distribution:** China (Guangdong, Hongkong).

17) *Bambusa multiplex* 'Solida'

**Local names:** Komachi-dake, Houbi-chiku (Japan)

**Synonyms:** *Bambusa multiplex* f. *solida*

*Bambusa glaucescens* f. *solida*

**Citations:** *Bambusa multiplex* 'Solida', Ohrnb., The Bamb. World. 269. 1999.——*B. multiplex* f. *solida* Muroi & I. Maruyama in Sugimoto, 1961.——*B. glaucescens* f. *solida* (Muroi & I. Maru- yama) Muroi & Sugimoto ex Muroi & H. Okamura, Take sasa, 149, 69. 1977.

**Characteristics:** Culms 3-5 m tall, 1-1.5 cm in diameter, internodes solid or subsolid; leaves 1-9 mm long, curly or curly at the apex.

**Use:** This bamboo can be planted for ornamentation.

**Distribution:** Japan (the middle part, southern islands of Honshu, part of western seacoast, northern Kyushu, southern Shikoku); China (central part).

18) *Bambusa multiplex* 'Stripestem Fernleaf'

**Local names:** Beni-houou-chiku (Japan); Stripestem Femleaf Hedge Bamboo (English)

**Synonyms:** *Bambusa floribunda* f. *viridi-striata*

*Bambusa glaucescens* cv. Stripestem Fernleaf

*Barnbusa glaucescens* f. *viridistriata*

*Bambusa multiplex* f. *stripestem fernleaf*

*Bambusa multiplex* 'Stripestem'

*Bambusa multiplex* f. *viridistriata*

*Bambusa multiplex* var. *elagans* f. *viridi-striata*

*Bambusa multiplex* var. *stripestemfernleaf*

*Bambusa nana* f. *viridi-striata*

*Bambusa nana* var. *typica* f. *viridi-striata*

*Leleba floribunda* f. *viridi-striata*

**Citations:** *Bambusa multiplex* 'Stripestem Fernleaf', Ohrnb., The Bamb.

World. 267. 1999; American Bamboo Society. *Bamboo Species Source List* No. 33: 8. Spring 2013.——*B. multiplex* (Lour.) Raeusch. cv. Stripestem Fernleaf R. A. Young in USDA Agr. Handb. No. 193: 41. 1961; Fl. Taiwan 5: 757. 1978.——*B. multiplex* 'Stripestem', R. A. Young ex Lin in Bull. Taiwan For. Res. Inst. No. 271, 1976:44.——*B. multiplex* var. *stripestemfernleaf* R. A. Young in Nation. Hort. Mag. 25: 261. 1946.——*B. multiplex* var. *elegans* (Koidz.) Muroi f. *viridi-striata* (Makino ex Tsuboi) Muroi ex Sugimoto, New Keys Jap. Trees 457. 1961; S. Suzuki, Index Jap. Bamb., 1978: 104, 340.——*B. multiplex* (Lour.) Raeuschel ex J. A. et J. H. Schult. f. *stripestem fernleaf* (R. A. Young) Yi in Yi et al. Icon. Bamb. Sin. 130. 2008, et in Clav. Gen. Spec. Bamb.Sin. 39. 2009; Shi et al. in The Ornamental Bamb. in China. 290. 2012.——*B. multiplex* f. *viridistriata* Muroi & Sugimoto ex Muroi in J. Himeji Gakuin Wom. Coll. No. 1, 1974.——*Bambusa glaucescens* 'Stripestem Fernleaf', Hatusima, Woody Pl. Jap., 1976:316.——*B. glaucescens* (Willd.) Sieb. ex Munro cv. Stripestem Fernleaf (R. A. Young) Chia et But in Phytologia 52(4): 259. 1982.——*B. glaucescens* f. *viridistriata* (? Makino ex Tsuboi) Muroi & Sugimoto, 1971: 10.——*B. nana* Roxb. var. *typica* Makino ex Tsuboi f. *viridi-striata* Makino ex Tsuboi, Ill. Jap. Sp. Bamb. ed. 2: 44. pl. 45. 1916.——*B. nana* f. *viridi-striata* Makino in Journ. Jap. 1: 28. 1917.——*B. floribunda* (Buse) Zoll. et Maur. ex Steud. f. *viridi-striata* Nakai in Riko Kyoiku 15; 66. 1932.——*Leleba floribunda* (Buse) Nakai f. *viridi-striata* Nakai in Journ. Jap. Bot. 9: 12. 1933.

**Characteristics:** Culms 1-3 m tall, initially light red and gradually becoming yellow with green stripes; branchlets pendulous, leaves 12-20, 1.6-3.8cm long.

**Use:** This bamboo can be planted for bonsai or ornamentation in parks, yards, and gardens.

**Distribution:** China (Hongkong, Taiwan); Japan; Europe; USA.

19) *Bambusa multiplex* 'Tiny Fern'

**Local names:** Tiny Fern Bamboo (English)

**Synonyms:** *Bambusa glaucescens* 'Tiny Fern'

**Citations:** *Bambusa multiplex* 'Tiny Fern', Ohrnb., The Bamb. World. 267. 1999; American Bamboo Society. *Bamboo Species Source List* No. 33: 9. Spring

2013. ——B. glaucescens 'Tiny Fern', Haubrich, 1981: 10.

**Characteristics:** Resemble *Bambusa multiplex* 'Fernleaf'. Culms 0.6 - 0.9 m; leaves slender, shorter than 2.5 cm.

**Use:** This bamboo can be planted for bonsaior ornamentation.

**Distribution:** USA.

20) *Bambusa multiplex* 'Variegata'

**Local names:** Hou-shiyou-chiku, Taiho-chiku (Japan);Silverstripe Hedge Bamboo (English)

**Synonyms:** *Bambusa argentea* var. *vittata*

*Bambusa glaucescens* 'Variegata'

*Bambusa glaucescens* f. *variegata*

*Barnbusa multiplex* f. *variegata*

*Bambusa multiplex* f. *vittato-argentea*

*Bambusa nana* var. *argenteostriata*

*Bambusa nana* var. *normalis* f. *vittato-argentea*

*Barnbusa nana* var. *normalis* f. *vittatoargentea*

*Barnbusa nana* var. *vatiegata*

*Bambusa scriptionis*

*Bambusa vittato-argentea*

*Leleba multiplex* f. *variegata*

**Citations:** *Bambusa multiplex* 'Variegata', Ohrnb., The Bamb. World. 268. 1999. ——*B. multiplex* f. *variegata* (Camus) R.A. Young ex A. V. Vasirev, 1956:29. ——*B. multiplex* f. *vittato-argentea* Nakai in Rika Kyo-iku 15 (6), 1932:67.——*B. glaucescens* 'Variegata' , Hatusima, Woody Pl. Jap., 1976:316. ——*B. glaucescens* f. *variegata* (Camus) Muroi & Sugimoto, 1971 : 10. ——*B. nana* var. *argenteostriata* hort. ex R.A. Young in Nation. Hort. Mag. 25, 1946: 260. ——*B. nana* var. *normalis* f. *vittato-argentea* Makino in S. Honda, Descr. Prod. For. Jap., 1900: 37. ——*B. nana* var. *normalis* f. *vittatoargentea* Makino ex Tsuboi, Illus. Jap. Sp. Bamb., 1916: 45, pl. XLVll. ——*B. nana* var. *vatiegata* Camus, Bamb., 1913: 12. ——*B. scriptionis* hort. ex W. Watson, 1889: 299. ——*B. vittato-argentea* hort. ex Mitford, Bamb. Gard.,

1896: 55, 216.——*B. argentea* var. *vittata* Beadle; R. A. Young in Nation. Hort. Mag. 25, 1946: 260. ——*Leleba multiplex* f. *variegata* (Camus) Nakai in J. Jap. Bot. 9, 1933:16.

**Characteristics:** Internodes with narrow creamy or white stripes; culm-sheaths with white stripes when fresh; leaves with creamy or white stripes as well.

**Use:** This bamboo can be planted for ornamentation.

**Distribution:** Japan; USA; Europe; Australia.

21) *Bambusa multiplex* 'Willowy'

**Local names:** Chuizhizhu (China); Willowy Hedge Bamboo (English)

**Synonyms:** *Bambusa multiplex* f. *willowy*

**Citations:** *Bambusa multiplex* 'Willowy', R. A. Young in Nation. Hort. Mag. 25, 1946: 260, 266; American Bamboo Society. *Bamboo Species Source List* No. 33: 9. Spring 2013; Ohrnb., The Bamb. World. 268. 1999; ——*B. multiplex* (Lour.) Raeusch. cv. Willowy R. A. Young, in USDA Agr. Handb. No.193: 42. 1961; Keng et Wang in Flora Reip. Pop. Sin. 9(1): 113. 1996. ——*B. multiplex* (Lour.) Raeuschel ex J. A. et J. H. Schult. f. *willowy* (R. A. Young) Yi in Yi et al. Icon. Bamb. Sin. 131. 2008, et in Clav. Gen.Spec. Bamb.Sin. 40. 2009; Shi et al. in The Ornamental Bamb. in China. 291. 2012.

**Characteristics:** Branches and leaves pendulous, leaves slender, 10-20 cm long, 8-16 mm wide.

**Use:** This bamboo can be planted for ornamentation.

**Distribution:** Europe; China (Guangzhou of Guangdong, Yibin of Sichuan); USA.

22) *Bambusa multiplex* 'Yellowstripe'

**Local names:** Huangwenzhu (China)

**Synonyms:** *Bambusa multiplex* f. *yellowstripe*

*Bambusa glaucescens* cv. Yellowstripe

**Citations:** *Bambusa multiplex* 'Yellowstripe', Ohrnb., The Bamb. World. 267. 1999. ——*B. glaucescens* (Willd.) Sieb. ex Munro cv. Yellowstripe Chia et C. Y. Sia in Guihaia 8(1): 57. 1988; Keng et Wang in Flora Reip. Pop. Sin. 9(1): 111. 1996. ——*B. multiplex* (Lour.) Raeuschel ex J. A. et J. H. Schult. f. *yellowstripe*

(Chia et C. Y. Sia) Yi in Yi et al. Icon. Bamb. Sin. 132. 2008, et in Clav. Gen.Spec. Bamb.Sin. 39. 2009.

**Characteristics:** Culms with yellow stripes on the branching side.

**Use:** This bamboo can be planted for ornamentation.

**Distribution:** China (Chengdu).

**(6) *Bambusa pervariabilis* McClure**

Culms 7-10 m tall, 4-5.5 cm in diameter, the top erect; internodes 30 cm long, with white powder initially and setae, the basal internodes with yellow-green stripes; a ring of grey-white tomenta above and below basal sheath-nodes. Branching from the basal first node, several to many branches clustered on nodes, 3 dominant branches strong and long. Culm-sheaths deciduous, thin leathery, glabrous or with a few setae abaxially, with yellow-green strips when fresh, the top asymmetrically arched; auricles unequal, wrinkled, oral setae undulate, the larger one extended downward, 3.5-4 cm long, 1 cm wide, the smaller one nearly round or oval, 1.5 cm long, 0.8 cm wide; ligules 3-4 mm tall, margins lobed, fimbriate; blades erect, deciduous, narrowly oval, with yellow-green stripes and brown setae abaxially, the base narrowed roundly and then linked with auricles, the linkage 3-7 mm, the base 2/3 as wide as the top of culm-sheaths. Branchlets with (4) 5 (6) leaves; margins of leaf sheaths with cilia; auricles obovate to obovate-oblong, oral setae present; ligules 0.5 mm tall; blades linear-lanceolate, 10-15 cm long, 1-1.5 cm wide, with dense pubescence abaxially, secondary veins 5-6 pairs, transverse veins inconspicuous. Pseudospikelets several clustered on flowering branches, linear, 2-5 cm long; 5-10 florets for each pseudospikelet, subtended by 2 or 3bracts with buds; rachilla 4 mm long; glume 1, oblong, 6 mm long, glabrous, 9 veins, apex acute; lemma oblong-lanceolate, 12-14 mm long, glabrous, 13-15 veins, apex acute; palea as long as lemma or a little shorter, 2 keels with cilia at the top, 6 veins between keels, 3 veins outside each keel; lodicules 3, unequal, margins with long cilia, the former two asymmetrically, 2.7 mm long, the back one a little bigger, obovate-oblong, 3 mm long; filaments short, anthers 5 mm long; ovary oblong, 1 mm long, the top with pubescence, style 1 mm long with pubescence, stigmas 3, 3 mm long, pubescent. Caryopsis broadly oblong, 1.5 mm long, the top with pubescence and residua of style and stigmas. Shooting from June to July.

**Use:** Culms can be used for building, furniture, farming tools, and so on; bamboo shavings are used for medicine.

**Distribution:** China (Guangdong, Guangxi, Fujian, Sichuan).

1) *Bambusa pervariabilis* 'Multistriata'

**Local names:** Punting Pole Bamboo (English)

**Synonyms:** *Bambusa pervariabilis* var. *multistriata*

**Citations:** *Bambusapervariabilis* 'Multistriata', Shi et al. in World Bamb. Ratt. 14(6): 27. 2016.——*B. pervariabilis* McClure var. *multistriata* W. T. Lin in Journ. Bamb. Res. 16 (3): 25. 1997;Yi et al. Icon. Bamb. Sin. 136. 2008, et in Clav. Gen.Spec. Bamb.Sin. 38. 2009; Shi et al. in The Ornamental Bamb. in China. 287. 2012.

**Characteristics:** Culms and culm-sheaths with white stripes; a ring of white tomenta absent within intranodes and below sheath-nodes.

**Use:** Culms can be used for building, furniture, farming tools, and so on; bamboo shavings are used for medicine.

**Distribution:** China (Guangxi, Guangzhou of Guangdong, Dujiangyan of Sichuan, Hongkong).

2) *Bambusa pervariabilis* 'Viridi-striata'

**Synonyms:** *Bambusa pervariabilis* var. *viridi-striata*

*Bambusa pervariabilis* var. *viridistriata*

**Citations:** *Bambusa pervariabilis* 'Viridi-striata', American Bamboo Society. *Bamboo Species Source List* No. 33: 9. Spring 2013.——*B. pervariabilis* McClure var. *viridi-striata* Q. H. Dai et X. C. Liu in Yi et al. Icon. Bamb. Sin. 136. 2008, et in Clav. Gen.Spec. Bamb.Sin. 38. 2009; Shi et al. in The Ornamental Bamb. in China. 287. 2012.——*B. pervariabilis* var. *viridistriata*, Ohrnb., The Bamb. World. 273. 1999.

**Characteristics:** Culms yellow with green stripes.

**Use:** This bamboo can be planted for ornamentation.

**Distribution:** China (Nanning and Rongshui of Guangxi).

***(7) Bambusa rigida* Keng et Keng f.**

Culms 7-12 m tall, 3.5-6 cm in diameter; internodes 30-45 cm long, with thin white powder initially, glabrous, a ring of grey-white tomenta above the basal first sheath-node. Branching from the basal third and fourth nodes, many branches clustered on nodes, dominant branches strong and long. Culm-sheaths deciduous, glabrous, sometimes the inner sides with procumbent and dark-brown setae at the base, the top asymmetrically arched; auricles unequal, a little wrinkled, oral setae undulate, the larger one ovate, 2.5 cm long, 1.5 cm wide, the smaller one ovate or round, 2/3 as large as the larger one; ligules 2.5-3 mm tall, lobed, fimbriate; blades erect, deciduous, ovate-triangular to ovate-lanceolate, with sparse brown setae abaxially, the base with dense brown setae adaxially, the base narrowed roundly and then extended laterally to link with auricles, the linkage 3-4 mm long, 2/5 as wide as the top of culm-sheaths, the basal margins with cilia. Leaf-auricles oblong, oral setae several; ligules 0.5 mm tall; blades 7.5-18 cm long, 1-2 cm wide, glabrous adaxially or with a few pubescence at the base, with dense pubescence abaxially. Pseudospikelets light green, single or several clustered on nodes of flowering branches, florets sterile when several pseudospikelets clustered, floret fertile when pseudospikelet single, fertile pseudospikelet 3-4.5 cm long, florets 3-7, the base subtended by several bracts with buds; rachilla flat, glabrous, 2-4 mm long, the top swollen like cup; glumes ovate, 6-7 mm long, veins many, apex acute; lemma oblong-lanceolate, 1-1.5 cm long, 4-8 mm wide, veins many, the middle vein prominent as keel, apex with a cusp; palea shorter than lemma, 2 keels with cilia at the upper part, 5 veins between keels; lodicules 3, 1.5-3 mm long, the upper margins with long cilia, the former 2 semi-spoon-like, the back one a little longer, obovate-lanceolate; anthers 4-6 mm long, apex with cilia, ovary 3-ridged, ovate, petiolate, 2-2.5 mm long together with petiole, the top with setae, style pubescent, 1.5-2 mm long, stigmas 3, pubescent, less than 1 mm.

**Use:** Culms are used for scaffolds and farming tools. This bamboo can be planted along riversides, on the hills, and near villages for ecological function.

**Distribution:** China (Sichuan, northern Guizhou, northeastern and southeastern Yunnan, Guangzhou of Guangdong, Xiamen of Fujian).

1) *Bambusa rigida* 'Luteolo-striata'

**Synonyms:** *Bambusa rigida* f. *luteolo-striata*

**Citations:** *Bambusa rigida* 'Luteolo-striata', Shi et al. in World Bamb. Ratt. 14(6): 27. 2016.——*B. rigida* Keng et Keng f. f. *luteolo-striata* Yi et L. Yangin Journ. Sichuan For. Sci. Techn. 36(2):24. 2015.

**Characteristics:** Culms and culm-sheaths green with light yellow stripes, branches sometimes with light yellow stripes as well.

**Use:** Culms are used for scaffolds and farming tools. This bamboo can be planted along riversides, on the hills, and near villages for ecological function and ornamentation.

**Distribution:** China (Changning of Sichuan).

### (8) *Bambusa textilis* McClure

Culms 8-10 m, 3-5 cm in diameter; internodes 40-70 cm, with white powder and light-brown setae when young, culm-walls 2-3 mm. Branching from the seventh to eleventh basal nodes, branches clustered, one dominant strong and long. Culm-sheaths deciduous, the base with dark-brown setae abaxially, the top asymmetrically broadly arched; auricles small, unequal, oral setae undulate, the larger one narrow-oblong, a little inclining, 1.5 cm long, 4-5 mm wide, the smaller one oblong, 1/2 as large as the larger one; ligules 2 mm tall, lobed, with short cilia; blades erect, deciduous, narrowly ovate-triangular, 2/3 as long as culm-sheaths or longer, the abaxial base with dark-brown setae, with setae or coarse among veins adaxially, the base narrowed as heart-shape, 2/3 as wide as the top of culm-sheaths. Leaf auricles present, falcate, oral setae radiate and undulate; ligules short, lobed; blades 9-17 cm long, 1-2 cm wide, with dense pubescence abaxially. Pseudospikelets single or several clustered on nodes, dark-purple when fresh and bronze when dry, linear-lanceolate, 3-4.5 cm long, 5-8 mm wide; prophyll broadly ovate, 3 mm long, 2 keels; 2 or 3 bracts with buds inside, broadly ovate, 3-4.5 mm long, glabrous, apex acute; 5-8 florets for each pseudospikelet, the top floret sterile;

rachilla semi-cylinder or flat, 4 mm long, the top swollen; glume 1, broadly ovate, 6 mm long, glabrous, 21 veins, apex acute; lemma oval, 11-14 mm long, glabrous, 25 veins, apex acute; palea lanceolate, 12-14 mm long, a little longer than lemma, 2 keels, 10 veins between keels, 4 veins outside each side of keels; lodicules unequal, margins with long cilia, the former 2 spoon- like, 3 mm long, the back one obovate-lanceolate, 2 mm long; filaments slender, anthers yellow, 5 mm long; ovary broadly ovate, 2 mm in diameter; the top thickened with setae, petiolate, style 0.7 mm long with setae, stigmas 3, 6-7 mm long, plumose.

1) *Bambusa textilis* 'Textilis'

**Local names:** Wong chuk, Weaver's Bamboo (English)

**Synonyms:** *Bambusa textilis* var. *textilis*

**Citations:** *Bambusa* 'Textilis', Shi et al. in World Bamb. Ratt. 14(6): 27. 2016.——*B. textilis* McClurevar. *textilis*, Keng et Wang in Flora Reip. Pop. Sin. 9(1): 124. 1996.

**Characteristics:** The same with *Bambusa textilis*.

**Use:** Culms are used for weaving. Some products from the internodes are used for medicine.

**Distribution:** China (southwestern, eastern, and central parts); USA.

2) *Bambusa textilis* 'Dwarf'

**Citations:** *Bambusa textilis* 'Dwarf', American Bamboo Society. *Bamboo Species Source List* No. 33: 10. Spring 2013.

**Characteristics:** Resemble *Bambusa textilis*. Culms short, 5-5.5 m tall, 3-3.3 cm in diameter.

**Use:** This bamboo can be planted for ornamentation and weaving.

**Distribution:** Unknown.

3) *Bambusa textilis* 'Glabra'

**Local names:** Smooth Weaver's Bamboo (English)

**Synonyms:** *Bambusa textilis* var. *glabra*

**Citations:** *Bambusa textilis* 'Glabra', Shi et al. in World Bamb. Ratt. 14(6): 27. 2016.——*B. textilis* McClure var. *glabra* McClure in Lingnan. Univ. Sci. Bull.

No. 9: 16. 1940; Keng et Wang in Flora Reip. Pop. Sin. 9(1):125.1996; Ohrnb., The Bamb. World. 276. 1999; Flora of China 22: 30. 2006; Yi et al. in Icon. Bamb. Sin. 143. 2008, et in Clav. Gen.Spec. Bamb.Sin. 44. 2009.

**Characteristics:** Internodes glabrous initially; culm-sheaths glabrous, blades 1/2 as long as the sheath or a little longer, the base roundly shrinked, ligules 1-1.5 mm tall.

**Use:** Culms are used for weaving. Some products from the internodes are used for medicine.

**Distribution:** China (Guangdong, Guangxi, Fujian, Hongkong).

4) *Bambusa textilis* 'Gracilis'

**Local names:** Slender Weaver's Bamboo (English)

**Synonyms:** *Bambusa textilis* var. *gracilis*

**Citations:** *Bambusa textilis* 'Gracilis', Shi et al. in World Bamb. Ratt. 14(6): 27. 2016.——*B. textilis* McClure var. *gracilis* McClure in Lingnan. Univ. Sci. Bull. No. 9: 16. 1940; Keng et Wang in Flora Reip. Pop.Sin.9(1):126. 1996; Ohrnb., The Bamb. World. 276. 1999; Flora of China 22: 30. 2006; Yi et al. in Icon. Bamb. Sin. 144. 2008, et in Clav. Gen.Spec. Bamb.Sin. 44. 2009;Shi et al. in The Ornamental Bamb. in China. 297. 2012.

**Characteristics:** Culms thinner, usually less than 3 cm; culm-sheaths with sparse brown setae at lateral sides and the base abaxially, blades 1/2 as long as the sheaths or a little shorter, the base slightly shrinked, ligules 1 mm tall.

**Use:** This bamboo are usually cultivated in parks, yards, and gardens for ornamentation.

**Distribution:** China (Guangdong, Guangxi, Fujian, Sichuan).

5) *Bambusa textilis* 'Kanapaha'

**Citations:** *Bambusa textilis* 'Kanapaha', American Bamboo Society. *Bamboo Species Source List* No. 33: 10. Spring 2013.

**Characteristics:** Culms large, 15-15.2 m tall, 6-6.4 cm in diameter, the lower part of the culm without branches and a little blue.

**Use:** This bamboo can be cultivated for ornamentation.

**Distribution:** USA.

6) *Bambusa textilis* 'Maculata'

**Synonyms:** *Bambusa textilis* f. *maculata*

*Bambusa textilis* var. *maculata*

**Citations:** *Bambusa textilis* 'Maculata', Ohrnb. in The bamboos of the world. 276. 1999; American Bamboo Society. *Bamboo Species Source List* No. 33: 10. Spring 2013.—— *B. textilis* McClurecv. *maculata* Chia et al. in Guihaia 8(2): 127. 1988; Keng et Wang in Flora Reip. Pop.Sin.9(1):125.1996. ——*B. textilis* McClure f. *maculata*(McClure) Yi in Journ. Sichuan For. Sci. Techn. 28(3): 17. 2007; Yi et al. in Icon. Bamb. Sin. 144. 2008, et in Clav. Gen.Spec. Bamb.Sin. 43. 2009; Shi et al. in The Ornamental Bamb. in China 295. 2012.——*B. textilis* McClure var. *maculata* McClure in Lingnan Univ. Sci. Bull. No. 9: 16. 1940.

**Characteristics:** Basal internodes and culm-sheaths with purple stripes.

**Use:** This bamboos is suitable for ornamentation in parks, yards, and gardens.

**Distribution:** China (Guangxi, Guangzhou of Guangdong).

7) *Bambusa textilis* 'Mutabilis'

**Citations:** *Bambusa textilis* 'Mutabilis', American Bamboo Society. *Bamboo Species Source List* No. 33: 10. Spring 2013.

**Characteristics:** Resemble *Bambusa textilis* 'Maculata'. Culms larger, 12-12.2 m tall, 5-5.8 cm in diameter, internodes long.

**Use:** This bamboo can be cultivated for ornamentation.

**Distribution:** Unknown.

8) *Bambusa textilis* 'Purpurascens'

**Synonyms:** *Bambusa textilis* f. *purpurascens*

*Bambusa textilis* var. *purpurascens*

**Citations:** *Bambusa textilis* 'Maculata', Ohrnb. in The bamboos of the world. 276. 1999. —— *B. textilis* McClure cv.Purpurascens Chia et al. in Guihaia 8(2): 127. 1988; Keng et Wang in Flora Reip. Pop. Sin. 9(1):125.1996. ——*B. textilis* McClure f. *purpurascens* (N. H. Xia) Yi in Journ. Sichuan For. Sci. Techn. 28(3): 17. 2007; Yi et al. in Icon. Bamb. Sin. 144. 2008, et in Clav. Gen.Spec. Bamb.

Sin. 43. 2009;Shi et al. in The Ornamental Bamb. in China. 296. 2012.——*B. textilis* McClure var. *purpureascens* N. H. Xia in Bamb. Res. 1985(1): 38. f. 2. 1985.

**Characteristics:** Internodes green with purple stripes in different size, sometimes the whole culm purple.

**Use:** This bamboos is suitable for ornamentation in parks, yards, and gardens.

**Distribution:** China (Guangdong, Fujian, Sichuan, Yunnan).

9) *Bambusa textilis* 'Scranton'

**Citations:** *Bambusa textilis* 'Mutabilis', American Bamboo Society. *Bamboo Species Source List* No. 33: 10. Spring 2013.

**Characteristics:** Culms sparsely clustered, culms 9-9.1 m tall, 5-5.1cm in diameter; branches short.

**Use:** This bamboo can be cultivated for ornamentation.

**Distribution:** Unknown.

10) *Bambusa textilis* 'Viridistriata'

**Synonyms:** *Bambusa textilis* f. *viridi-striata*

**Citations:** *Bambusa textilis* 'Viridistriata', Shi et al. in World Bamb. Ratt. 14(6): 27. 2016.——*B. textilis* McClure f. *viridi-striata* Yi in Journ. Bamb. Res. 21(1): 10. 2002; Yi et al. in Icon. Bamb. Sin. 146. 2008, et in Clav. Gen.Spec. Bamb.Sin. 43. 2009; Shi et al. in The Ornamental Bamb. in China. 295. 2012.

**Characteristics:** Internodes yellow with green stripes; culm-sheaths green with yellow stripes when fresh.

**Use:** This bamboo can be cultivated for ornamentation.

**Distribution:** China (Guangxi, Sichuan).

**(9) *Bambusa tulda* Roxb.**

Culms 8-10 m, 5-7 cm in diameter; internodes 40-45 cm, with white powder when young, aerial roots and a ring of grey tomenta above basal sheath-nodes. Branching from the basal first node, many branches clustered, dominant branches strong and long. Culm-sheaths deciduous, thick-leathery, with densely dark-brown procumbent setae and white powder abaxially when young, the top asymmetrically arched, margins with short cilia; auricles unequal, undulately wrinkled, the larger one extended downward, 1/3 as long as the top of culm-sheaths, kidney-shaped or obovate-lanceolate, 4.5-5 cm long, 1.5 cm wide; ligules 1.5-2 mm tall, entire; blades erect, broadly ovate, the base heart-shaped or roundly narrowed, then extended to link with auricles, the linking parts 1-1.3 cm long, coarse or with setae adaxially, the base 5/8 as wide as the top of culm-sheaths, the basal margins undulate with cilia. Leaf auricles undeveloped or absent, oral setae 1-2 or absent; ligules lobed; blades 15-20 cm long, 1.5-2.5 wide, the base with setae adaxially sometimes, light-green and with dense pubescence abaxially. Pseudospikelets single or 2-5 clustered on nodes, linear or linear-lanceolate, 2.5-7.5 cm long, 5 mm wide, 4-6 florets for each pseudospikelet, the upper 1 to 2 florets sterile; rachilla rod-like, the surface toward palea flat and with stripes, apex ciliate; glumes 1-2, veins many, apex acute; lemma ovate to oblong, 1.2-2.5 cm long, 7.5 mm wide, glabrous, veins many, the apex acute, margins with several cilia; palea shorter than lemma, 2 keels with cilia, apex with pubescence, 5-7 veins between keels; lodicules 3, 3.8 mm long, the former two thickened at base and with 5 veins, margins with long cilia; anthers purplish, 7.5 -10 mm long, apex blunt or a little concaved; ovary obovate or obovate-oblong, the top thickened and with long setae, style short with setae, stigmas 3, plumose. Caryopsis oblong, 7.5 mm long, hilum groove-shaped on adaxial surface, apex with long setae.

**Distribution:** China (Guangdong, Guangxi, southern Tibet, Fujian, Sichuan).

1) *Bambusa tulda* 'Striata'

**Citations:** *Bambusa tulda* 'Striata', American Bamboo Society. *Bamboo Species Source List* No. 33: 10. Spring 2013.

**Characteristics:** Culms 21.3 m tall, 10.2 cm in diameter; culms with yellow stripes, more at the base.

**Use:** This bamboo can be cultivated for ornamentation.

**Distribution:** Unknown.

**(10) *Bambusa tuldoides* Munro**

Culms 6-10 m, 3-5 cm in diameter; internodes 30-36 cm, with white powder initially, glabrous; a ring of grey-white tomenta above and below sheath-nodes of the basal first and second nodes. Branching from the basal first or second node, dominant branches strong. Culm-sheaths deciduous, glabrous, with 1-3 yellow-white stripes at one lateral margin initially, the top asymmetrically arched; auricles unequal, the larger one ovate or ovate-oblong, 2.5 cm long, 1-1.4 cm wide, a little wrinkled, oral setae slender, the smaller one ovate or oblong, half as large as the larger one, ligules 3-4 mm, fimbriate; blades erect, deciduous, asymmetrically ovate-triangular to narrowly triangular, with brown setae on both surfaces when young, the base narrowed and then extened to auricles, 2/3 to 3/4 as wide as the top of culm-sheaths, the basal margins wrinkled, oral setae undulate. Leaf auricles absent or present; ligules truncate, short; blades 10-18 cm, 1.5-2 cm wide, glabrous or pubescent at the base adaxially, with dense pubescence abaxially, the base round or broadly wedge-shaped. Pseudospikelets clustered on nodes and subtended by bracts, light-green, flat, linear-lanceolate, 2-3 cm long, 3-4 mm wide; prophyll with 2 keels, keels with cilia; 2 bracts with buds inside, glabrous, apex blunt; 6 or 7 florets for each pseudospikelet, the upper and lower florets sterile; rachilla flat, 3-4 mm long, the top cup-shaped with pubescence; glume 1, ovate-oblong, 8.5 mm long, glabrous, apex acute; lemma ovate-oblong, 11-14 mm long, 19 veins, glabrous, apex blunt; palea as long as lemma or a little shorter than lemma, keels with white cilia, both 4 veins for inter-keels and outside keels, apex blunt with a cluster of white pubescence; lodicules 3, obovate, margins with long cilia; the former 2 asymmetric, broad and short, 2.5 mm long, the back one narrow, 3 mm long; anthers 3 mm long; ovary obovate, 1.2 mm long, petiolate, the top thickened with long and tough pubescence; style 0.7 mm long, with setae, stigmas 3, 5.5 mm long, plumose. Caryopsis cylindrical, 8 mm long, 1.5 mm in diameter, the apex blunt with setae and the residua of the style. Shooting from July to September.

1) *Bambusa* 'Tuldoides'

**Local names:** Shuizhu, Yingshengtaozhu, Yingsantaozhu, Yingtouhuangzhu (China); Buloh balai (Malaysia); Bambu blenduk (Indonesia); Verdant Bamboo, Punting Pole Bamboo (English)

**Citations:** *Bambusa* 'Tuldoides', Shi et al. in World Bamb. Ratt. 14(6): 27. 2016. ——*B. tuldoides* Munro cv.Tuldoides,Keng et Wang in Flora Reip. Pop. Sin. 9(1): 87. 1996.

**Characteristics:** The same with *Bambusa tuldoides*.

**Use:** Culms are used for buildings, furniture, and farming tools. The bamboo shavings are used for medicine.

**Distribution:** China (Guangdong, Hongkong, Guangxi, southern Guizhou, Fujian, Yunnan, Taiwan); Vietnam; Southeast Asia; South America.

2) *Bambusa tuldoides* 'Swolleninternode'

**Local names:** Gujieqingganzhu; Butto-chiku (Japan); Buddha Bamboo (English)

**Synonyms:** *Bambusa tuldoides* f. *swolleninternode*

*Bambusa tuldoides* 'Ventricosa'

**Citations:** *Bambusa tuldoides* 'Swolleninternode', Xia in Bamb. Res. No. 23, 1985: 38, fig.1; Shi et al. in World Bamb. Ratt. 14(6): 28. 2016.——*B. tuldoides* Munro cv. Swolleninternode N. H. Xia in Bamb. Res. 1985 (1): 38. f. 1. 1985; Keng et Wang in Flora Reip. Pop.Sin.9(1):88.1996.——*B. tuldoides* Munro f. *swolleninternode* (N. H. Xia) Yi in Journ. Sichuan For. Sci. Techn. 28(3): 17. 2007; Yi et al. in Icon. Bamb. Sin. 147. 2008, et in Clav. Gen.Spec. Bamb.Sin. 38. 2009; Shi et al. in The Ornamental Bamb. in China. 288. 2012. ——*B. tuldoides* 'Ventricosa', Ohrnb., The Bamb. World. 277. 1999.

**Characteristics:** Basal internodes shortened and swollen as the drum shape.

**Use:** This bamboo are planted for ornamentation in parks and yards, or as bonsai.

**Distribution:** China (Guangdong, Fujian); Southeast Asia; USA; Europe.

3) *Bambusa tuldoides* 'Ventricosa Kimmer'

**Local names:** Kimmei-daifuku-chiku, Kimmei-butto-chiku (Japan)

**Synonyms:** *Bambusa tuldoides* f. *kimmei*

**Citations:** *Bambusa tuldoides* 'Ventricosa Kimmer', Ohrnb., The Bamb. World. 278. 1999.

**Characteristics:** Culms yellow with several green stripes, grooves light green; leaves with several white stripes.

**Use:** This bamboo Can be planted for ornamentation in parks and yards, or as bonsai.

**Distribution:** Japan.

### (11) *Bambusa ventricosa* McClure

Normal culms to 10 m tall, 5 cm in diameter, the top a little pendulous, the base zigzag; internodes 30-35 cm long, glabrous, a ring of grey-white tomenta above and below sheath-nodes, culm-walls 6-12 mm thick; sheath-nodes glabrous. Branching from basal third or fourth node, branches 1-3, branchlets of them sometimes shortened into soft thorns, middle and upper nodes with many branches, 3 dominant, strong and long. Abnormal culms to 5 m tall, thinner than normal culms, internodes shortened and swollen like bottles, to 6 cm long; branch solitary, internodes swollen. Culm-sheaths deciduous, ribs prominent, glabrous, the top nearly symmetrically broadly arched or truncate; auricles unequal, oral setae present, the larger one 5-6 mm wide, the smaller one 3-5 mm wide; ligules 0.5-1 mm tall, fimbriate; blades erect, ovate or ovate-lanceolate, the base narrowed heart-shaped, a little narrower than the top of culm-sheaths. Branchlets with (5) 7-11 leaves; auricles ovate or falcate, oral setae present; ligules nearly truncate; blades 9-18 cm long, 1-2 cm wide, with dense pubescence abaxially. Pseudospikelets solitary or several clustered on nodes, linear-lanceolate, flat, 3-4 cm long; prophyll broadly ovate, 2.5-3 mm long, 2 keels with cilia, apex blunt; bracts 1 or 2 with buds, narrowly ovate, 4-5 mm long, veins 13-15, apex acute; 6-8 florets fertile for each pseudospikelet, basal 1 or 2 and top 2 or 3 Sterile; rachilla flat, 2-3 mm long, the top swollen like a cup, with cilia; glume absent or 1, ovate-elliptical, 6.5-8 mm long, veins 15-17, apex acute; lemma glabrous, ovate-elliptical, 9-11 mm long, veins 19-21, with transverse veins, apex acute; palea as long as lemma, 2 keels with cilia at the top, 4 veins between and outside keels, apex tapering, apex with a cluster of white cilia; lodicules 3, 2 mm long, upper margins with long cilia, the former two a little asymmetrical, the back one broadly elliptical; filaments slender, anthers yellow, 6 mm long, apex blunt; ovary petiolate, broadly ovate, 1-1.2 mm long, the top thickened with pubescence, style very short with pubescence, stigmas 3, 6 mm long, plumose.

1) *Bambusa ventricosa* 'Nana'

**Local names:** Xiaofoduzhu, Fozhu, Huluzhu（China）

**Synonyms:** *Bambusa tuldoides* 'Nana'

**Citations:** *Bambusa ventricosa* McClure cv. Nana Wen in Journ. Bamb. Res. 4(2): 18. 1985; Keng et Wang in Flora Reip. Pop. Sin. 9(1): 70. 1996.——*B. tuldoides* 'Nana', Ohrnb., The bamboos of the world. 278. 1999.

**Characteristics:** Culms shortened conspicuously like Buddha's belly.

**Use:** This bamboo are planted for ornamentation.

**Distribution:** China (cultivated in the pots or on the ground in souther China); Japan; Thailand; Vietnam.

2) *Bambusa ventricosa* 'Kimmei'

**Local names:** Jinsihuluzhu（China）;Kimmei-daifuku-chiku, Kimmei-butto-chiku (Japan)

**Synonyms:** *Bambusa ventricosa* f. *kimmei*

**Citations:** *Bambusa ventncosa* 'Kimmei', Muroi & Y. Tanaka ex H. Okamura in H. Okamura & Tanaka, 1986: 7. ——*B. ventricosa* McClure f. *kimmei* Muroi et Y. Tanaka ex H. Okamura & Y. Tanaka in Hort. Bamb. Sp. Jap., 7, 101, invalid (Engl. descr.). 1986; Yi et al. in Icon. Bamb. Sin. 112. 2008, Yi et al.in Clav. Gen.Spec. Bamb. Sin. 32. 2009; Shi et al. in The Ornamental Bamb. in China. 279. 2012.

**Characteristics:** Culms yellow, internodes with several green stripes on the branching side; leaves with light yellow-white stripes.

**Use:** This bamboo can be planted as bonsai.

**Distribution:** China (Nanjing of Jiangsu, Chengdu of Sichuan); Japan.

**(12) *Bambusa vulgaris* Schrader ex Wendland**

Culms 8-15 m tall, 5-9 cm in diameter, the base erect or a little zigzag; internodes 20-30 cm, initially with thin white powder and light-brown procumbent setae; intranodes of basal internodes with short aerial roots, with grey tomenta above and below sheath-nodes. Branching from the basal parts of culms, many branches clustered, dominant branches long and thick. Culm-sheaths deciduous, with dense brown-black setae abaxially, the top arched; auricles large, almost equal, oblong or kidney-shaped, upward, 8-10 mm wide, oral setae undulate; ligules 3-4 mm tall, fimbriate; blades erect or open, deciduous, broadly triangular or triangular, with sparse brown setae abaxially and dense brown setae adaxially, narrowed inside, 1/2 as broad as the top of culm-sheaths, the basal margins with setae. Leaf-sheaths with sparse brown setae initially; auricles broadly falcate when present, oral setae several or absent; ligules entire; blades 10-30 cm long, 1.3-2.5 cm wide, glabrous, the base nearly round, a little asymmetric. Pseudospikelets clustered on nodes, a little flat, narrowly or linearly lanceolate, 2-3.5 cm long, 4-5 mm wide, 5-10 florets for each pseudospikelet subtended by several bracts; rachilla 1.5-3 mm long; glumes 1 or 2, only the top parts with pubescence abaxially, apex acute; lemma 8-10 mm long, the top parts with pubescence abaxially, apex acute; palea shorter than lemma, 2-keeled with cilia; lodicules 3, 2-2.5 mm long, margins with long cilia; anthers 6 mm long, with clusters of short pubescence at the top; style long and slender, 3-7 mm long, stigmas 3, short.

1) *Bambusa* 'Vulgaris'

**Local names:** Taishanzhu (China); Bambu ampel (Indonesia); Buloh aur, Buloh pau, Buloh minyak, Aur beting (Malaysia);Mai-luang, Phai-luang (Thailand); Daisan-chiku (Japan); Gemeiner Bambus (Germany); Common Bamboo (English)

**Citations:** *Bambusa* 'Vulgaris', Shi et al. in World Bamb. Ratt. 14(6): 28. 2016.——*Bambusa vulgaris* Schrader ex Wendlandcv. Vulgaris,Keng et Wang in Flora Reip. Pop. Sin. 9(1): 96. 1996.

**Characteristics:** The same with *Bambusa vulgaris*.

**Use:** Culms are used for buildings, paper making, or farming.

**Distribution:** China (Yunnan, Guangxi, Guangdong, Hongkong, Fujian); tropical Asia; Madagascar.

2) *Bambusa vulgaris* 'Vittata'

**Local names:** Qingsijinzhu, Longtouzhu (China); Buloh gading, Aur gading,

Buloh kuning (Malaysia); Bambu kuning (Indonesia); KJnshi-chiku (Japan); Golden Common Bamboo (English)

**Synonyms:** *Bambusa striata*

*Bambusa vulgaris* f. *vittata*

*Bambusa vulgads* 'Striata'

*Bambusa vulgaris* var. *striata*

*Bambusa vulgaris* var. *vittata*

**Citations:** *Bambusa vulgaris* 'Vittata', Hatusima, Woody Pl. Jap., 1976: 315; Ohrnb. in The bamboos of the world. 279. 1999.; American Bamboo Society. *Bamboo Species Source List* No. 33: 11. Spring 2013.——*B. vulgaris* Schrader ex Wendland cv. Vittata (A. et C. Riv.) McClure in Agr. Handb. USDA. No. 193: 46. 1961; Keng et Wang in Flora Reip. Pop. Sin.9(1):97.1996.——*B. striata* Loddiges ex Lindley in Penny Cycl., 3, 1835:357.——*B. vulgaris* 'Vittata', McClure ap. Swallen in Fieldiana Bot. 24 (2), 1955:60. ——*B. vulgaris* var. *striata* (Loddiges ex Lindley) Gamble in Ann. Roy. Bot. Gard. Calcutta 7, 1896: 44, pl. 40 fig. 4-5. ——*B. vulgaris* Schrader ex Wendland f. *vittata* A. et C. Riv.1982: 467.——*B. vulgaris* Schrader ex Wendland f. *vittata*(A. et C. Riv.) Yi in Journ. Sichuan For. Sci. Techn. 28(3): 17. 2007; Yi et al. in Icon. Bamb. Sin. 149. 2008, et in Clav. Gen.Spec. Bamb. Sin. 36. 2009; Shi et al. in The Ornamental Bamb. in China. 283. 2012.——*B. vulgaris* Schrader ex Wendland var. *vittata* A. et C. Riv. in Bull. Soc. Acclim. Ⅲ 5: 640. 1878.

**Characteristics:** Culms yellow with green stripes, culm-sheaths green with yellow stripes when fresh.

**Use:** This bamboo can be planted for ornamentation in parks, yards, and scenic area.

**Distribution:** China (Fujian, Taiwan, Guangdong, Guangxi, Hongkong, Hainan, Yunnan); cultivated around the world (tropical and subtropical areas of East Asia, South Asia, Southeast Asia, Madagascar).

3) *Bambusa vulgaris* 'Wamin'

**Local names:** Bambu blenduk (Indonesia);Wamin Bamboo (English)

**Synonyms:** *Bambusa vulgaris* cv. Waminii

*Bambusa vulgaris* f. *wamin*

*Bambusa wamin*

**Citations:** *Bambusa vulgaris* 'Wamin', Ohrnb. in The bamboos of the world. 280. 1999; American Bamboo Society. *Bamboo Species Source List* No. 33: 11. Spring 2013.——*B. vulgaris* Schrader ex Wendland cv. WaminMcClure, Keng et Wang inFlora Reip. Pop. Sin. 9(1): 97. 1996.——*B. vulgaris* Schrader ex Wendland f. *waminii* Wen in Journ. Bamb. Res. 4(2): 16. 1985; Yi et al. in Icon. Bamb. Sin. 150. 2008, et in Clav. Gen.Spec. Bamb.Sin. 35. 2009; Shi et al. in The Ornamental Bamb. in China. 282. 2012. ——*B. wamin* Brandis ex Camus, Bamb., 1913: 135.

**Characteristics:** Culms green or light yellow-green, basal internodes extremely shortened and swollen.

**Use:** This bamboo is of great value for ornamentation.

**Distribution:** China (southern China, Zhejiang, Fujian, Taiwan, southwestern Sichuan, southern Yunnan); Southeast Asia; USA; Europe.

4) *Bambusa vulgaris* 'Wamin Striata'

**Citations:** *Bambusa vulgaris* 'Wamin Striata', American Bamboo Society. *Bamboo Species Source List* No. 33: 11. Spring 2013.

**Characteristics:** Resemble *Bambusa vulgaris* 'Wamin'. Culms light green with dark green stripes.

**Use:** This bamboo can be cultivated for ornamentation.

**Distribution:** Unknown.

## 2.2 *Chimonobambusa* Makino

### (1) *Chimonobambusa angustifolia* C. D. Chu & C. S. Chao

Culms 2-5 m tall, 1-2 cm in diameter, terete, or basal internodes a little quadrate, with dense white pubescence and sparse setae, coarse after the setae deciduous; sheath-nodes with light brown cilia initially; culm-nodes flat or prominent on the branching nodes; basal intranodes with aerial roots 9-14.

Branches 3 on each node, or more than three, branches solid, nodes extremely prominent. Culm-sheaths shorter than internodes, papery to thickly papery, yellow-brown, the upper part with pale grey or light yellow spots abaxially, the blower part with sparse light yellow pubescence and setae, margins with yellow brown cilia, ribs conspicuous, transverse veins purple; ligules truncate or arched, entire, ciliate; blades tiny, tapering triangular, 3-5 mm long, auricles absent or inconspicuous. Branchlets with leaves 1-3 (4); auricles absent or inconspicuous, oral setae several, 3-5, erect, 3-5 mm long; ligules short; blades linear lanceolate or linear, 6-15 cm long, 0.5-1.2 cm wide, secondary veins 3-4 pairs. Flowers unknown. Shooting from August to September.

1) *Chimonobambusa angustifolia* 'Repleta'

**Synonyms:** *Chimonobambusa angustifolia* f. *repleta*

**Citations:** *Chimonobambusa angustifolia* C. D. Chu et C. S. Chao f. *repleta* Yi et H. R. Qi in Journ. Bamb. Res. 23(3): 6. 2004; Yi et al. in Icon. Bamb. Sin. 266. 2008, et in Clav. Gen.Spec. Bamb.Sin.79. 2009.

**Characteristics:** Culms shorter, 1.8 m tall, 0.5 cm in diameter; culms and branches solid.

**Use:** This bamboo can be cultivated for ornamentation.

**Distribution:** China (Liangping of Chongqing).

**(2) *Chimonobambusa marmorea* (Mitford) Makino**

Rhizomes amphipodium. Culms diffuse or a little caespitose, 1-1.5 (3) m tall, 0.5-1 cm in diameter; internodes terete or flat and grooved on the branching side, 10-14 cm long; sheath-nodes with brown tomenta initially; culm-nodes a little prominent; basal intranodes with aerial roots. Branches 3. Culm-sheaths persistent, slightly longer than internodes, yellow brown with pale spots abaxially, glabrous, or with sparse light yellow setae, margins ciliate; auricles absent; ligules truncate or a little arched; blades tapering, erect, 2-3 mm long. Leaves 2-3 for each branchlets; oral setae 3-4 mm long; blades 10-14 cm long, 0.7-0.9 cm wide, glabrous, secondary veins 4-5 pairs. Flowering branches subtended by a serial of bracts gradually larger, the middle and upper with 1-4 pseudospikelets;

pseudospikelets linear, 2-4 cm long, bracts 0-2 with or without buds; 4-7 florets; rachilla 3-4 mm long, glabrous; glumes 1-2 or absent, 6-8 mm long, veins 5-7; lemma green or a little purple, 6-7 mm long, veins 5-7; palea as long as lemma, 2-keeled, two veins between keels, apex truncate or bifid, two veins outside each keel; lodicules 3; stamens 3, filaments free; ovary slender ovate, style one, short, stigmas 2, plumose. Caryopsis with thick pericarp, nut-like when dry.

**Use:** This bamboo can be cultivated for ornamentation, especially as bonsai.

**Distribution:** China (Zhejiang, Fujian, Sichuan, Yunnan, Guangxi); Japan; Europe; USA.

1) *Chimonobambusa marmorea* 'Gimmei'

**Local names:** Gimmei-kan-chiku (Japan)

**Synonyms:** *Chimonobambusa marmorea* f. *gimmei*

**Citations:** *Chimonobambusa marmorea* 'Gimmei', Ohrnberger, Bamb. World. Chimonobambusa ed. 2, 1996:18; Ohrnb., The Bamb. World. 181. 1999. ——*Ch. marmorea* (Mitford) Makino f. *gimmei* Muroi et Kasahara in Rep. Fuji Bamb. Gard. No. 17: 8, 1972; Muroi in J. Himeji Gakuin Wom. Coll. No. 1,1974: 2; H. Okamura & al., Il1. Hort. Bamb. Sp. Jap., 1991 : 349; Yi et al. in Icon. Bamb. Sin. 272. 2008, et in Clav. Gen. Spec. Bamb. Sin.76. 2009.

**Characteristics:** Culms diffuse or a little caespitose, 1-1.5 (3) m tall, 0.5-1 cm in diameter; culms green with light yellow green stripes on the groove; internodes terete, 10-14 cm long; sheath-nodes with brown tomenta initially; culm-nodes a little prominent; basal intranodes with several aerial roots. Branches 3. Culm-sheaths persistent, a little longer than internodes, brown, with pale spots, glabrous, or with sparse light yellow setae, margins ciliate; auricles absent; ligules truncate or arched; blades tapering, erect, 2-3 mm long. Leaves 2-3 for each branchlet; blades with white stripes; oral setae 3-4 mm long; blades 10-14 cm long, 0.7-0.9 cm wide, glabrous, secondary veins 4-5 pairs. Caryopsis with thick pericarp, nut-like when dry.

**Use:** This bamboo can be cultivated for ornamentation, especially as bonsai.

**Distribution:** Japan; China; Europe; USA.

2) *Chimonobambusa marmorea* 'Kimmei'

**Local names:** Suisho-chiku, Somoku-kinyoshyu (Japan)

**Synonyms:** *Chimonobambusa marmorea* f. *kimrnei*

**Citations:** *Chimonobambusa marmorea* 'Kimmei', Ohrnberger, Bamb. World Gen. Chimonobambusa, 1990:26; Ohrnb., The Bamb. World. 182. 1999.——*Ch. marmorea* f. *kimmei* Muroi & H. Okamura in J. Himeji Gakuin Wom. Coll. No. 1, 1974: 2.

**Characteristics:** Resemble *Chimonobambusa marmorea* 'Gimmei'. Culms with golden stripes.

**Use:** This bamboo can be cultivated for ornamentation.

**Distribution:** Japan.

3) *Chimonobambusa marmorea* 'Variegata'

**Local names:** Chigo-kan-chiku (Japan); meaning small winter bamboo, Chiryo-kan-chiku, Beni-kan-chiku, Heisaku-kan-chiku (English)

**Synonyms:** *Arundinaria marmorea* 'Variegata'

*Arundinaria marmorea* var. *variegata*

*Chimonobambusa marrnorea* f. *albovatiegata*

*Chimonobambusa marmorea* f. *variegata*

*Chimonobambusa marmorea* var. *variegata*

**Citations:** *Chimonobambusa marmorea* 'Variegata', Ohwi, Fl. Jap. 2nd Ed., 1965:135; Ohrnb., The Bamb. World. 181. 1999.——*Ch. marmorea* (Mitford) Makino f. *variegata* (Makino) Ohwi in Fl. Jap., 75. 1953;Yi et al. in Icon. Bamb. Sin. 272. 2008, et in Clav. Gen. Spec. Bamb. Sin.76.2009; Shi et al. in The Ornamental Bamb. in China. 352. 2012. ——*Ch. marmorea* var. *variegata* (Maki-no) Makino in Bot. Mag. Tokyo 28, 1914: 154. ——*Ch. marrnorea* f. *albovatiegata* Rifat, Nouv. Tahiti, 24 Feb., 1986: 34. ——*Arundinaria marmorea* 'Variegata', A. H. Lawson, Bamb. Gard. Guide, 1968:157.——*A. marmorea* var. *variegata* Makino in S. Honda, Descr. Prod. For. Jap., 1900: 38; Makino in Bot. Mag.Tokyo 14, 1900: 63.

**Characteristics:** Resemble *Chimonobambusa marmorea* 'Gimmei'. Culms yellow, internodes with several green stripes; culms can become red from yellow under the sunshine; leaf blades with white stripes.

**Use:** This bamboo is usually cultivated for ornamentation.

**Distribution:** Japan; China (Nanjing of Jiangsu, Dujiangyan of Sichuan); Europe; USA.

### (3) *Chimonobambusa quadrangularis* (Fenzi) Makino

Culms 3-8 m tall, 1-4 cm in diameter; internodes 8-22 cm long, terete, or basal internodes a little quadrate, with brown verrucate setae initially, coarse when setae deciduous; sheath-nodes with brown tomenta and setae initially; culm-nodes flat or prominent on branching side; intranodes below the middle culm with aerial roots. Branches 3. Culm-sheaths deciduous, shorter than internodes, glabrous or with sparse procumbent setae, transverse veins purple, margins ciliate; auricles and ligules inconspicuous; blades tapering, 3-5 mm long. Leaves 2-5 for each branchlet, sheaths glabrous, oral setae erect; ligules glabrous, margins ciliate; blades oblong lanceolate, 9-29 cm long, 1-2.7 cm wide, pubescent abaxially, secondary veins 4-7 pairs. Flowering branches raceme or paniculate, terminal branches slender and glabrous, the base subtended by several gradually larger bracts, pseudospikelets 2-4, sometimes one pseudospikelet present at the base of the flowering branch with fewer bracts; pseudospikelets slender, 2-3 cm long, lateral pseudospikelets with prophyll; 2-5 florets for each pseudospikelet, sometimes one or two sterile florets present at the basal most; rachilla 4-6 mm long, glabrous; glumes 1-3, lanceolate, 4-5 mm long; lemma papery, green, lanceolate or ovate lanceolate, veins 5-7; palea as long as lemma; lodicules oblong ovate; anthers 3.5-4 mm long; stigmas 2, plumose.

1) *Chimonobambusa quadrangularis* 'Albostriata'

**Local names:** Fuiri-hôchiku, Fuiri-shikaku-dake (Japan)

**Synonyms:** *Chimonobambusa quadrangularis* f. *albostriatus*

*Tetragonocalamus quadrangularis* f. *albostriatus*

**Citations:** *Chimonobambusa quadrangularis* 'Albostriata', Ohrnberger, Bamb. World Gen. Chimonobambusa, 1990:38; J. P. Demoly in Bamb. Assoc.

Europ. Bamb. EBS Sect. Fr. No. 8:22, 1991; Ohrnb., The Bamb. World. 184. 1999. ——*Ch. quadrangularis* f. *albostriatus* Stover, Bamb. Book, 1983: 37. ——*Ch. quadrangularis* f. *albostriata* (Muroi & H. Okamura) Wen in J. Bamb. Res. 10 (1): 17, 1991.——*Tetragonocalamus quadrangularis* f. *albostriatus* Muroi & H. Okamura in Rep. Fuji Bamb. Gard. No. 17, 1972: 10; Muroi in J. Himeji Gakuin Wom. Coll. no. 1: 11, 1974.

**Characteristics:** Leaves with white stripes.

**Use:** This bamboo can be cultivated in gardens for ornamentation.

**Distribution:** Japan.

2) *Chimonobambusa quadrangularis* 'Aureostriata'

**Local names:** Kishima-hôchiku, Kishima-shika-ku-dake (Japan)

**Synonyms:** *Chirnonobambusa quadrangularis* f. *aureostriata*

*Tetragonocalamus quadrangularis* f. *albostriatus*

**Citations:** *Chimonobambusa quadrangularis* 'Aureostriata', Ohrnberger, Bamb. World Gen. Chimonobambusa, 1990:38; Ohrnb., The Bamb. World. 184. 1999. ——*Ch. quadrangularis* f. *aureostriata* (Muroi & H. Okamura) Wen in J. Bamb. Res. 10 (1) : 17, 1991.——*Tetragonocalamus quadrangularis* f. *albostriatus* Muroi & H. Okamura in Rep. Fuji Bamb. Gard. No. 17, 1972: 10; Muroi in J. Himeji Gakuin Wom. Coll. No. 1: 11, 1974.

**Characteristics:** Leaves with golden stripes.

**Use:** This bamboo can be cultivated for ornamentation.

**Distribution:** Japan.

3) *Chimonobambusa quadrangularis* 'Cyrano de Bergerac'

**Synonyms:** *Chirnonobambusa quadrangularis* f. *cyrano-berger-aca*

*Tetragonocalamus quadrangularis* f. *striatus*

**Citations: *Chimonobambusa quadrangularis*** 'Cyrano de Bergerac', Rifat in J. Bamb. Res. 6 (2): 25, 1987; Ohrnb., The Bamb. World. 184. 1999. ——*Ch. quadrangularis* f. *cyrano-berger-aca* (Rifat) Wen in J. Bamb. Res. 10 (1): 18, 1991. ——*Tetragonocalamus quadrangularis* f. *striatus* Rifat, ined., ex M. Hirsh, Europ. Bamb. Netw. Newsl. 3, 14, 1986.

**Characteristics:** Culms green with yellow stripes。

**Use:** This bamboo can be cultivated for ornamentation.

**Distribution:** Japan; Europe.

4) *Chimonobambusa quadrangularis* 'Joseph de Jussieu'

**Local names:** Kimmei-hôchiku, Kimmei-shikaku-dake (Japan)

**Synonyms:** *Chimonobambusa quadrangularis* f. *nagaminea*
*Chimonobambusa quadrangularis* f. *nagamineus*
*Chimonobambusa quadrangularis* 'Nagamine'
*Tetragonocalamus quadrangulans* f. *castillonis*
*Tetragonocalamus quadrangularis* 'Nagamineus'
*Tetragonocalamus quadrangularis* f. *nagamineus*

**Citations:** *Chimonobambusa quadrangularis* 'Joseph de Jussieu', Rifat in J. Bamb. Res. 6 (2) :25, 1987; Ohrnb., The Bamb. World. 185. 1999.——*Ch. quadrangularis* f. *nagaminea* (Muroi & H. Hamada) Wen in J. Bamb. Res. 10 (1): 18, 1991.——*Tetragonocalamus quadrangulans* f. *castillonis* Rifat, ined., ex M. Hirsh in Europ. Bamb. Netw. Newsl. 3, 1986:14.——*T. quadrangulans* 'Nagamineus', Muroi & H. Hamada; cf. H. Okamura in H. Okamura & Y. Tanaka, Hort. Bamb. Sp. Japo, 1986: 31.——*Tetragonocalamus quadrangularis* f. *nagamineus* Rifat, ined., ex M. Hirsh, Europ. Bamb. Netw. Newsl. 3, 14, 1986; Muroi & H. Hamada ex H. Okamura in H. Okamura & Y. Tanaka, Hort. Bamb. Sp. Jap., 1986: 31.fig. 30, and H. Okamura & M. Konishi in I. c., 1986:89; H. Okamura & al., III. Hort. Bamb. Sp. Jap., 1991:349.——*Ch. quadrangularis* 'Nagamine', Ohrnberger, Bamb. World Gen. Chimonobambusa, 1990: 40, based on *Tetragonocalamus quadrangularis* f. *nagamineus* Muroi & Hamada ex H. Okamura *Chimonobambusa quadrangularis* f. *nagaminea* G. Bol in Amer. Bamb. Soc. Newsl. 11 (3) : 3, 1990.

**Characteristics:** Culms golden with several green stripes, especially in the groove; leaves ocassionally with white stripes.

**Use:** This bamboo can be planted in yards, parks, and scenic area for ornamentation.

**Distribution:** Japan (southern Honshu, Kyushu, Kagoshima); China.

5) *Chimonobambusa quadrangularis* 'Purpureoculmis'

**Synonyms:** *Chimonobambusa quadrangularis* f. *purpureiculma*

**Citations:** *Chimonobambusa quadrangularis* (Fenzi) Makino f. *purpureiculma* Wen in Journ. Bamb. Res. 8(1): 24. 1989; Ohrnb., The Bamb. World. 186. 1999; Yi et al. in Icon. Bamb. Sin. 278. 2008, et in Clav. Gen.Spec. Bamb.Sin.79. 2009.

**Characteristics:** Culms purple.

**Use:** This bamboo can be planted in gardens; bamboo shoots are edible.

**Distribution:** China (Shunchang and Guixi of Fujian).

6) *Chimonobambusa quadrangularis* 'Sotaroana'

**Local names:** Gomafu-hôchiku, Gomafu-shikaku-dake (Japan)

**Synonyms:** *Chimonobambusa quadrangularis* f. *sotaroana*
  *Chimonobambusa quadrangularis* 'Napoleon-Bonaparte'
  *Tetragonocalamus quadrangularis* 'Sotaroana'
  *Tetragonocalamus quadrangularis* f. *sotaroanus*
  *Tetragonocalamus quadrangulans* vat. *sotaroanus*

**Citations:** *Chimonobambusa quadrangularis* 'Sotaroana', Ohrnberger, Bamb. World Gen. Chimonobambusa, 1990:38.; Ohrnb., The Bamb. World. 185. 1999.—— *Ch. quadrangularis* f. *sotaroana* (Muroi) Wen in J. Bamb. Res. 10 (1) : 17, 1991.—— *Ch. quadrangularis* 'Napoleon-Bonaparte', Rifat in J. Bamb. Res. 6 (2): 25, 1987—— *Tetragonocalamus quadrangularis* 'Sotaroana', Hatusima, Woody Pl. Jap., 1976. ——*T. quadrangularis* f. *sotaroanus* (Muroi) Muroi in J. Himeji Gakuin Worn. Coll. No. 1, 1974:11. ——*T. quadrangulans* vat. *sotaroanus* Muroi in Hyogo Biol. 2, 1948:7.

**Characteristics:** Culms yellow occasionally with several green stripes.

**Use:** This bamboo can be planted for ornamentation.

**Distribution:** Japan (southern Honshu); Europe (introduced from Japan to France and Switzerland by C. Rifat in 1987).

7) *Chimonobambusa quadrangularis* 'Suow'

**Local names:** Tatejima-hôchiku, Suow-shikaku- dake (Japan)

**Synonyms:** *Chimonobambusa quadrangularis* f. *suhow*
*Chimonobambusa quadrangularis* f. *suou*
*Chimonobambusa quadrangulans* f. *suow*
*Tetragonocalamus quadrangulans* 'Tatejima'
*Tetragonocalamus quadrangulans* f. *tatejima*

**Citations:** *Chimonobambusa quadrangularis* 'Suow', Ohrnb., The Bamb. World. 185. 1999.——*Ch. quadrangulans* f. *suow* (Kasahara & H. Okamura) Wen in J. Bamb. Res. 10 (1), 1991 : 18, invalid (basionym not validly published) *Chimonobambusa quadrangulans* 'Suow'; Ohrnberger, Bamb. World Gen. Chimonobambusa, 1990: 40. ——*Ch. quadrangularis* f. *suhow* G. Bol in Amer. Bamb. Soc. Newsl. 9 (6): 2, 1988. ——*Ch. quadrangularis* f. *suou* G. Bol in Amer. Bamb. Soc. Newsl. 11 (3) :3, 1990, invalid Chimonobambusa quadrangulans 'Suou'; G. Cooper in Amer. Bamb. Soc. Newsl. 16 (4) : 17, 1995.——*Tetragonocalamus quadrangularis* 'Suow', Kasahara & H. Okamura; cf. H. Okamura in H. Okamura & Y. Tanaka, Hort. Bamb. Sp. Jap., 1986: 31.——*T. quadrangularis* f. *suow* Kasahara & H. Okamura in H. Okamura & Y. Tanaka, Hort. Bamb. Sp. Jap., 1986: 31, fig. 31. ——*T. quadrangulans* 'Tatejima', Kasahara & H. Okamura; cf. H. Okamura in H. Okamura & Y. Tanaka, Hort. Bamb. Sp. Jap., 1986: 31.——*T. quadrangulans* f. *tatejima* Kasahara & H. Okamura ex H. Okamura & al., III. Hort. Bamb. Sp. Jap., 1991: 34,9.

**Characteristics:** Culms yellow with several green stripes; leaves sometimes with yellow and white stripes.

**Use:** This bamboo can be planted for ornamentation.

**Distribution:** Japan (found in southern Japan in 1980); USA (introduced from Japan by American Bamboo Society during 1988-1990).

8) *Chimonobambusa quadrangularis* 'Yellow Groove'

**Citations:** *Chimonobambusa quadrangularis* 'Yellow Groove', American Bamboo Society. *Bamboo Species Source List* No. 33: 13. Spring 2013.

**Characteristics:** Culms green with grooves yellow.

**Use:** This bamboo can be cultivated for ornamentation.

**Distribution:** Unknown.

**(4) *Chimonobambusa szechuanensis* (Rendle) Keng f.**

Culms 2.5-4 (6) m tall, 1.5-2 cm in diameter; internodes 18-22 cm long, terete, or basal internodes nearly quadrate, glabrous; sheath-nodes with sparse brown tomenta; culm-nodes flat or a little prominent; basal intranodes with aerial roots. Branches 3. Culm-sheaths tardily deciduous, shorter than internodes, glabrous, with purple stripes, margins ciliate; ligules 0.5-1 mm tall; blades tapering triangular, 3-5 mm long. Leaves 1-3 for each branchlet; oral setae 3-5 mm long; ligules 1-1.5 mm tall; blades 18-20 cm long, 1.2-1.5 cm wide, secondary veins 4-6 pairs. Flowering branches branching iteratively, terminal with or without leaves, branches and pseudospikelets mixed on the node, pseudospikelets 1-3, subtended by 0-4 bracts, the upper one or two bracts with buds or secondary pseudospikelets; florets 3 or 4; glumes 2 or 3; lemma ovate lanceolate, apex tapering, veins 7-9; palea oblong, as long as lemma, apex obtuse or concave, 2-keeled; lodicules 3, with two larger abaxially, membranous, upper margins ciliate; anthers yellow; ovary ovate, style extremely short, stigmas 2, plumose. Caryopsis oblong, 15 mm long, 6 mm thick, pericarp 0.8-1 mm thick, nut-like, pericarp and episperm connected.

1) *Chimonobambusa szechuanensis* 'Szechuanensis'

**Synonyms:** *Chimonobambusa szechuanensis* var. *szechuanensis*

**Citations:** *Chimonobambusa szechuanensis* (Rendle) Keng f. var. *szechuanensis*, Keng et Wang in Flora Reip. Pop. Sin. 9(1): 337. 1996.

**Characteristics:** The same with *Chimonobambusa szechuanensis*.

**Use:** This bamboo can be planted for ecological construction and ornamentation.

**Distribution:** China (western Sichuan and western Yunnan).

2) *Chimonobambusa szechuanensis* 'Flexuosa'

**Synonyms:** *Chimonobambusa szechuanensis* f. *flexuosa*

*Chimonobambusa szechuanensis* var. *flexuosa*

**Citations:** *Chimonobambusa szechuanensis* var. *flexuosa* Hsueh et C. Li in

Journ. Yunn. For. Coll. 1982(1): 40. f. 3. 1982; Hsueh et W. P. Zhang in Bamb. Res. 7(3): 10. 1988; Keng et Wang in Flora Reip. Pop. Sin. 9(1): 337. 1996; Yi et al. in Icon. Bamb. Sin. 279. 2008, et in Clav. Gen.Spec. Bamb.Sin.78. 2009.—— *Chimonobambusa szechuanensis* f. *flexuosa* (Hsueh et C. Li)Wen et Ohrnb., Gen. Chimonobambusa 44. 1990; Ohrnb., The Bamb. World. 187. 1999.

**Characteristics:** Basal internodes zigzag and shortened.

**Use:** This bamboo can be planted for ornamentation.

**Distribution:** China (Ya'an of Sichuan).

## 2.3 *Dendrocalamopsis* (Chia & H. L. Fung) Keng f.

### (1) *Dendrocalamopsis lineariaurita* Yi et L.Yang

Rhizomes sympodium. Culms 14-15 m tall, 5.5-8 cm in diameter, the top erect; internodes 35-42, 32-38 cm (42 cm) long, basal internodes 10-33 cm long, terete, glabrous, grey green, initially with thin white powder, hollow, culm-wall 5-12 mm thick, pith clastic; sheath-nodes prominent, brown, glabrous; culm-nodes flat, basal internodes with aerial roots; intranodes 4-12 mm tall, glabrous. Culm buds peach-shaped, margins ciliate. Branching from 4-6 m above the ground, branches many, dominant branches 2.5-3 m long, 1 cm in diameter, lateral branches slender, 2-4 mm in diameter, spreading. Culm-sheaths deciduous, leathery, slightly shorter than internodes, with brown setae and white powder abaxially, margins without cilia; auricles linear, 2-4 mm wide, grey; ligules arched, grey, 2-3 mm tall, margins ciliate; blades triangular or long triangular, erect, 10-25 cm long, 6-8 cm wide, with brown setae abaxially, margins serrate. Leaves 5-9 for each branchlet; sheaths glabrous, the upper with thin white powder, margins without cilia; auricles absent, oral setae purple brown or grey brown, to 2 mm long; ligules arched, purple brown, 1 mm tall, margins ciliate to 2 mm long; petiole 2-3 mm long, light green, glabrous; blades linear or linear lanceolate, green, papery, glabrous, 15-25 cm long, 1.7-3.2 cm wide, the base truncate to wedge-shaped, apex tapering, secondary veins 8-9 pairs, transverse veins oblong, margins serrate on one side, nearly smooth on the other side. Shooting in August.

**Use:** This bamboo can be planted for ecological construction and ornamentation.

**Distribution:** China (Dujiangyan of Sichuan).

1) *Dendrocalamopsis lineariaurita* 'Luridilineata'

**Synonyms:** *Dendrocalamopsis lineariaurita* f. *luridilineata*

**Citations:** *Dendrocalamopsis lineariaurita* Yi et L.Yang f. *luridilineata* Yi et L. Yang in Journ. Sichuan For. Sci. Techn. 36(3):3. Fig. 6. 2015.

**Characteristics:** Basal internodes with light yellow stripes.

**Use:** This bamboo is cultivated in gardens for ornamentation.

**Distribution:** China (Dujaingyan of Sichuan).

### (2) *Dendrocalamopsis oldhami* (Munro) Keng f.

Rhizomes sympodium. Culms caespitose, 6-12 m tall, 3-9 cm in diameter; internodes 20-35 cm long, a little zigzag, with white powder when young, culm-walls 4-12 mm thick. Branching from upper nodes, branches many, clustered, dominant branches 3, strong and long. Culm-sheaths deciduous, the top truncate, glabrous or with sparse or dense brown setae, margins without cilia or with cilia at upper part; auricles nearly equal, elliptical or nearly round, oral setae present; ligules 1 mm tall, entire or undulate; blades erect, triangular, the base truncate and narrowed, 1/2 as wide as the top of culm-sheaths. Branchlets with leaves 6-15; sheaths with setae initially; auricles semicircle, oral setae brown; ligules short; blades 15-30 cm long, 3-6 cm wide, pubescent abaxially, secondary veins 9-14 pairs, transverse veins conspicuous, margins coarse or with bristles. Flowering branches without leaves; the base of pseudospikelets green, upper part red purple, flat laterally, 2.7-3 cm long, 7-10 mm wide, single or clustered on nodes; bracts 3-5, the upper 1 or 2 without buds inside; 5-9 florets for each pseudospikelet; rachilla disarticulated under glumes; glume 1, ovate, 9-10 mm long, 8 mm wide, margins with cilia, veins many, transverse veins present; lemma ovate, 17 mm long, 13 mm wide, glabrous or pubescent, veins 31, transverse veins present, margins with cilia; palea 13 mm long, pubescent on both surfaces, apex acute, 2 keels, 3-5 veins between keels, 2 veins outside each keel, transverse veins present between veins, margins and keels with cilia; lodicules 3, ovate-lanceolate, 3.5 mm long, veins conspicuous, margins with cilia; stamens 6, filaments free, anthers 8 mm long; ovary ovate, 2 mm long, with setae, stigmas 3, plumose. Shooting from May to November. Flowering in summer and autumn.

1) *Dendrocalamopsis oldhami* 'Oldhami'

**Local names:** Nizhu, Shizhu, Maolvzhu, Wuyaozhu, Changzhizhu, Xiaojiaolv.

**Synonyms:** *Dendrocalamopsis oldhami* f. *oldhami*

**Citations:** *Dendrocalamopsis oldhami* (Munro) Keng f. f. *oldhami*, Keng et Wang in Flora Reip. Pop. Sin. 9(1): 141. 1996.

**Characteristics:** The same with *Dendrocalamopsis oldhami*.

**Use:** Bamboo shoots are edible and delicious, and they can be eaten freshly or processed. The shooting period is long. Culms are used for building, weaving, or paper making. The bamboo shavings are used for medicine.

**Distribution:** China (Zhejiang, Fujian, Taiwan, Guangdong, Guangxi, Hainan).

2) *Dendrocalamopsis oldhami* 'Revoluta'

**Synonyms:** *Bambusa oldhami* f. revoluta

*Dendrocalamopsis oldhami* f. *revoluta*

*Neosinocalamus revolutus*

*Sinocalamus oldhamii* f. *revolutus*

**Citations:** *Dendrocalamopsis oldhami* (Munro) Keng f. f. *revoluta* (W. T. Lin et J. Y. Lin) W. T. Lin in Guihaia 10 (1): 15. 1990; Keng et Wang in Flora Reip. Pop. Sin. 9(1): 142. 1996; Yi et al. in Icon. Bamb. Sin. 181. 2008, et in Clav. Gen. Spec. Bamb.Sin.54. 2009.——*Bambusa oldhami* Munro f. *revoluta* W. T. Lin et J. Y. Lin in Act. Phytotax. Sin. 26 (3): 224. f. 2. 1988; Ohrnb., The Bamb. World. 272. 1999.——*Neosinocalamus revolutus* (W. T. Lin & J. Y. Lin) Wen in J. Bamb. Res. 10 (1), 1991: 23. ——*Sinocalamus oldhamii* f. *revolutus* (W. T. Lin & J. Y. Lin) W. T. Lin in J. S. China Agr. Univ. 14 (3), 1993:111.

**Characteristics:** Culms green with yellow stripes; intranodes with a ring of pale grey or light yellow tomenta. The top of culm-sheaths wide, yellow brown setose abaxially at the base; auricles oblong, reflexed, margins ciliate; ligules 1.5 mm tall, serrate; blades triangular, the base a little extended outward to link with auricles. Branching from the basal third node. Lemma without transverse veins; scales nearly ovate; style very short.

**Use:** Bamboo shoots are edible and delicious, and they can be eaten freshly or processed. The shooting period is long. Culms are used for building, weaving, or paper making. The bamboo shavings are used for medicine.

**Distribution:** China (Guangdong, Zhejiang).

3) *Dendrocalamopsis oldhami* 'Striata'

**Synonyms:** *Dendrocalamopsis oldhami* f. *striata*

**Citations:** *Dendrocalamopsis oldhami* (Munro) Keng f. f. *striata* Y. Y. Wang et W. Y. Zhang in Journ. Bamb. Res. 25(1): 26. 2006; Yi et al. in Icon. Bamb. Sin. 181. 2008, et in Clav. Gen. Spec. Bamb. Sin.54. 2009.

**Characteristics:** Culms light yellow with green stripes. Culm-sheaths green with yellow stripes. Some leaves with yellow stripes.

**Use:** This bamboo is usually planted in parks and scenic area for ornamentation. Bamboo shoots are edible; culms are used for building, weaving, or paper making.

**Distribution:** China (Ruian of Zhejiang).

# 3   Publications on Bamboo Cultivar Registration

Formal publications related to international cultivar registration for bamboos during 2015-2016.

1) SHI J Y, JIN X B, 2015. The Establishment and Progress of the International Cultivar Registration Authority for Bamboos. Cultivated Plant Taxonomy News, (3):12-13.

2) SHI J Y, WANG D Y, YI T P, MA L S, ZHANG X L, YAO J, 2015. A New Dendrocalamus rongchengensis Cultivar 'Hualongdan'. Forest Research, 28(3): 441-442.

3) SUN M S, SHI J Y, YI T P, MA L S, YAO J, PU Z Y, 2015. A New Chimonocalamus delicates Cultivar 'Hongyun'. Acta Hortic. Sin, 42 (12):2555-2556.

4) SHI J Y et al., 2015. International Cultivar Registration Report for Bamboos ( 2013-2014). Beijing: Science Press: 1-110.

5) YAO J, SUN M S, SHI J Y, ZHOU D Q, PU Z Y, YANG Z J, 2016. A New Cultivar 'Jindian Huazhu' from *Phyllostachys vivax*. Science and technology information, (7): 97-98.

6) SHI J Y, ZHOU D Q, MA L S, YAO J, PU Z Y, 2016. The Directional Breeding and Feasibility of Functional Bamboos. Agricultural Science & Technology, 17(3): 711-716.

7) SHI J Y, DAI M L, ZHOU D Q, YAO J, GAO G C, 2016. A New Cultivar 'Meiling' from Neosinocalamus fangchengensis. World Bamboo and Rattan, 14(5):31-33, 45.

8) SHI J Y, YI T P, ZHOU D Q, MA L S, YAO J, 2016. Theory and Practice of the International Bamboo Cultivars Registration. World Bamboo and Rattan, 16 (6): 23-28.

9) SUN M S, SHI J Y, ZHOU D Q, YAO J, PU Z Y, 2016. A New Bamboo Cultivar 'Qioushi' from *Dendrocalamus membranaceus*. North Garden, (24):157-159.

# 4 Website Construction relevant to international registration for bamboo cultivars

## 4.1 Name and Website
- Name: International Cultivar Registration Center for Bamboos (ICRCB)
- Website: http://www.bamboo2013.org or http://www.icrcb.org

## 4.2 Website languages
Chinese and English.

## 4.3 Contents of the website
- *Home page*: General introduction to the ICRCB, information updating, quick-entry for the newly registered cultivars, and hyperlinks to websites of the International Society for Horticultural Science and other relevant organizations.
- ICRCB *introduction*: it includes textual and pictorial display.
- *Information updating*: ICRCB releases timely progresses and news of new bamboo cultivar registration.
- *Cultivar registration*: It issues information of the cultivars published recently including descriptions and photos.
- *Research achievements*: It exhibits significant works and publications in the field of international bamboo cultivar registration.
- *Professional team*: It contains brief introductions to the current experts of the International Cultivar Registration Committee for Bamboos.
- *Registration gardens*: It introduces the current International Bamboo Cultivar Registration Gardens.
- *Downloads*: It provides downloads of ICRCB internal information (such

as the application form, the range of international registration for bamboo cultivars) and free downloads for the references in the field of international registration for bamboo cultivars.

- *Contact*: It provides the address (such as query of online digital maps), contacting telephone number, postal code, email address and a "dialogue pannel".

## 4.4 Website operation

The website was opened to the public in October 2014. In July 2015, the website was upgraded and revised, which includes:

- Based on the Chinese version, English version was added, which means the website has begun on internationalization.
- The website was redesigned with more beautiful appearance, which made the photos more clear.
- Mobile web service was launched, which made it possible for the users to visit the website anywhere and anytime.
- A commentry entry was opened in order that the users can issue their own comments, reviews, opinions and suggestions. It is good for two-way interaction.
- A co-sharing column was launched for the domestic and oversea users to share their ideas and information via facebook, Twitter, Microblog, Wechat and Tencentspace.
- Online application entry can be found in the website. The users can submit their online application for the new cultivars.

The website employs technicians to ensure normal, 24 hours full-time run in good service. The website now is not only to release information and display functions, but also to provide individuals or organizations both at home and abroad with internet service.

## 4.5 Website Service

Since the website opened to the public, it has provided the individuals, organizations and social bodies with internet service. The users cover the research institutions, universities, enterprises, governmental departments, NGOs and individuals. In 2015-2016, the website hits were more than 10,000 times. And downloads reached 1,500 times plus 200 times hyperlinks by the website homepage.

Facing the growing users and visitors, the next step for the website is to open the user registration service. The website will classify the users into the browsing users, real-name registration users and offline users to provide different high-quality service to various users. We are going to open the public number service special for the Wechat users in Chinese version in order to convert passive serivice into active service with the aim at providing timely registration and release information for bamboo cultivar international registration.

# 5 Construction of International Cultivar Registration Garden for Bamboos

**5.1 International Cultivar Registration Garden for Bamboos (Beijing, China)**

International Cultivar Registration Garden for Bamboos (Beijing, China), in abbreviation of ICRGB (Beijing, China) with the registration number IC-001-2013-001. It is located in Beijing, the capital city of China, established in October 2013 with 5 hm$^2$ land size. Its objective is to collect, keep, register and directionally breed hardy ornanmental bamboo cultivars from the North China. The garden has not yet operated due to some reasons.

**5.2 International Cultivar Registration Garden for Bamboos (Chengdu, China)**

International Cultivar Registration Garden for Bamboos (Chengdu, China), in abbreviation of ICRGB (Chengdu, China) with the registration number IC-001-2014-002. It is located in Chengdu City, the capital city of Sichuan Province, China, established in June 2014 with 12.5 hm$^2$ land size. Its objective is to collect, keep, register and directionally breed ornamental bamboo cultivars under the various weather conditions from Sichuan Basin and the regions with the same weather conditions. Since the garden was established, five new bamboo cultivars have been registered. They are listed as follows:

1) *Neosinocalamus affinis* 'Doupengzhu'
2) *Neosinocalamus affinis* 'Foducizhu'
3) *Neosinocalamus affinis* 'Niutuizhu'
4) *Neosinocalamus affinis* 'Shetouzhu'
5) *Dendrocalamus rongchengensis* 'Hualongdan'

**5.3 International Cultivar Registration Garden for Bamboos (Dujiangyan, China)**

International Cultivar Registration Garden for Bamboos (Dujiangyan, China), in abbreviation of ICRGB (Dujiangyan, China) with the registration number IC-001-2014-003. It is located in Dujiangyan City, Sichuan Province, China, established in August 2014 with 15 hm$^2$ land size. Its objective is to collect, keep, register and directionally breed bamboo cultivars for shoots under the various weather conditions from Sichuan Basin and the regions with the same weather conditions. Since the garden was established, two new bamboo cultivars have been registered. They are listed as follows:

1） *Chimonobambusa neopurpurea* 'Dujiangyanfangzhu'

2） *Chimonobambusa neopurpurea* 'Lineata'

## 5.4　International Cultivar Registration Garden for Bamboos (Kunming, China)

International Cultivar Registration Garden for Bamboos (Kunming, China), in abbreviation of ICRGB (Kunming, China) with the registration number IC-001-2016-004. It is located in Kunming City, the capital city of Yunnan Province, China, established in July 2016 with 2 hm$^2$ land size. Its objective is to collect, keep, register and directionally breed ornamental bamboo cultivars from the highland in Kunming region and the regions with the same weather conditions. Since the garden was established, two new bamboo cultivars have been registered. They are listed as follows:

1） *Chimonocalamus delicates* 'Caiyun'

2） *Chimonocalamus delicates* 'Hongyum'

## 5.5　International Cultivar Registration Garden for Bamboos (Nanyang, China)

International Cultivar Registration Garden for Bamboos (Nanyang, China), in abbreviation of ICRGB (Nanyang, China) with the registration number IC-001-2016-005. It is located in Nanyang City, Henan Province, China, established in December 2016 with 10 hm$^2$ land size. Its objective is to collect, keep, register and directionally breed ornamental bamboo cultivars from Nanyang and the regions with the same weather conditions. Since the garden was established, One new bamboo cultivar has been registration:

1） *Neosinocalamus fangchengensis* 'Meiling'

# 主要参考文献
# Selected Bibliography

陈松河，张万旗，包宇航，等，2015. 厦门市竹笋业发展现状及对策［J］. 现代农业科技，22：316-318.

柯启柱，2013. 矮脚麻竹笋高产栽培技术［J］. 长江蔬菜，14：50，51.

雷霆，江明艳，刘一颖，等，2015. 新品种竹海硬头黄选育试验初报［J］. 竹子研究汇刊，34（4）：28-31.

靳晓白，成仿云，张启翔，2013. 国际栽培植物命名法规［M］. 北京：中国林业出版社.

史军义，易同培，马丽莎，等，2012. 中国观赏竹［M］. 北京：科学出版社.

史军义，马丽莎，2014. 竹类国际栽培品种登录的原则与方法［J］. 林业科学研究，27（2）：246-249.

史军义，王道云，易同培，等，2015. 龙丹竹新品种'花龙丹'［J］. 林业科学研究，28（3）：441，442.

史军义，2015. 国际竹类栽培品种登录报告（2013-2014）［M］. 北京：科学出版社.

史军义，代梅灵，周德群，等，2016. 方城慈竹一新品种'美菱'［J］. 世界竹藤通讯，14（5）：31-33，45.

史军义，易同培，周德群，等，2016. 国际竹类栽培品种登录的理论与实践［J］. 世界竹藤通讯，16（6）：23-29，41.

孙茂盛，史军义，易同培，等，2015. 香竹一新品种'红云'［J］. 园艺学报，42（12）：2555，2556.

孙茂盛，史军义，周德群，等，2016. 黄竹新品种'秋实'的选育［J］. 北方园艺，24：157-159.

姚俊，孙茂盛，史军义，等，2016. 乌哺鸡竹一新品种'金殿花竹'［J］. 科技信息，7：97，98.

易同培，史军义，马丽莎，等，2008. 中国竹类图志［M］. 北京：科学出版社.

易同培，李本祥，2012. 佯黄竹——我国经济竹子一新种［J］. 四川林业科技，33（3）：7-10.

易同培，史军义，马丽莎，等，2014. 刺黑竹一新变型及棉花竹的一新异名［J］. 四川林业科技, 35（1）: 18-20.

易同培, 杨林, 史军义, 等, 2015. 绿竹属新分类群和另一竹子新变型 [J]. 四川林业科技, 36（3）: 1-4.

易同培, 史军义, 马丽莎, 等, 2016. 河南慈竹属一新种 [J]. 四川林业科技, 37（2）: 1-3.

中国科学院中国植物志编辑委员会, 1996. 中国植物志第九卷第一分册 [M]. 北京: 科学出版.

American Bamboo Society, 2013. Bamboo species source list No. 33 [M]. Spring.

BRICKELL C D, ALEXANDER C, DAVID J C, et al., 2016. International Code of Nomenclature for Cultivated Plants [S]. 9 th ed. The International Union of Biological Sciences International Commission for the Nomenclature of Cultivated Plants. Leuven: ISHS.

LI D Z, WANG Z P, ZHU Z D, et al., 2006. Flora of China Vol. 22 (Poaceae, Tribe Bambuseae) [M]. Science Press, Beijing & Missouri Botanical Garden Press, St. Louis.

MCNEILL J, BARRIE F R, BUCK W R, et al., 2012. International Code of Nomenclature for algae, fungi, and plants (Melbourne Code) [S]. The Eighteenth International Botanical Congress Melbourne, Australia. Königstein: Koeltz Scientific Books.

OHRNBERGER D, 1999. The bamboos of the world [M]. Amsterdam: Elsevier.

SHI J Y, JIN X B, 2015. The Establishment and Progress of The International Cultivar Registration Authority for Bamboos [J]. Cultivated Plant Taxonomy News, (3):12, 13.

SHI J Y, ZHOU D Q, MA L S, et al., 2016. The Directional Breeding and Feasibility of Functional Bamboos [J]. Agricultural Science & Technology,17(3): 711-716.